Principles and Practice of Variable Pressure/Environmental Scanning Electron Microscopy (VP-ESEM)

Principles and Practice of Variable Pressure/Environmental Scanning Electron Microscopy (VP-ESEM)

Debbie J Stokes

Published in association with the Royal Microscopical Society

Series Editor: Mark Rainforth

A John Wiley and Sons, Ltd, Publication

This edition first published 2008
© 2008 John Wiley & Sons Ltd

Registered office
John Wiley & Sons Ltd, The Atrium, Southern Gate, Chichester, West Sussex, PO19 8SQ, United Kingdom

For details of our global editorial offices, for customer services and for information about how to apply for permission to reuse the copyright material in this book please see our website at www.wiley.com.

Library of Congress Cataloging-in-Publication Data

Stokes, Debbie, J.
  Principles and practice of variable pressure/environmental scanning
electron microscopy (VP-ESEM) / Debbie J Stokes.
      p. cm.
  Includes bibliographical references and index.
  ISBN 978-0-470-06540-2 (cloth)
  1.   Scanning electron microscopy. I. Title.
  QH212.S3S76 2008
  502.8′25–dc22
                         2008029099

A catalogue record for this book is available from the British Library.

ISBN 978-0-470-06540-2

Typeset in 10.5/13 Sabon by Laserwords Private Limited Chennai, India
Printed and bound in Great Britain by TJ International, Padstow, Cornwall, UK

This book is dedicated to the memory of Ralph Knowles, who tragically died in June 2008 as this book was going to press.

Ralph played a major role in the development of many aspects of ESEM technology, from its commercial realization with Electroscan Corp., through the transition to FEI-Philips and FEI Company, where he had recently been appointed Director of R&D, North America. His technological achievements in various areas of electron microscopy were widely recognized and his many years of experience, dating back to his time at Cambridge Instruments in 1975, highly valued.

Ralph was an inspiration to many, all around the globe. I was fortunate enough to meet Ralph and others from Electroscan during the heady, early days of ESEM, all sharing 'the passion'. This led to more than a decade of Ralph's unerring support, mentoring and friendship. When considering whether to undertake writing this book, I immediately turned to Ralph for words of wisdom and guidance. His advice was to try to keep the details as generic as possible, to be accessible not only to the users of 'ESEM' instruments but to those of other manufacturers. His philosophy was this: all of these microscopes rely on a chamber gas for their operation, the rest is a just variation on the theme. That is why this book is firmly centered on the physical effects of performing electron microscopy in a gas environment and genuinely tries to avoid any commercial reference. This is a tribute to Ralph's personal integrity and reflects my own wish to help as many people as possible, as far as I am able.

Contents

Preface

Scanning electron microscopy (SEM) is a technique of major importance and is widely used throughout the scientific and technological communities. The modern scanning electron microscope is capable of imaging details of the order of tens of Ångstroms (i.e. sub-nanometre), subject to the limits of electron–specimen interactions. However, for a long time it has been apparent that the high-vacuum SEM needed to develop in respects other than increased resolution. Hence, the advent of SEMs that utilise a gas for image formation while simultaneously providing charge stabilisation for electrically nonconductive specimens. Some instruments also allow for thermodynamic stabilisation of hydrated specimens. These microscopes are known by terms such as 'environmental', 'extended pressure' and 'variable pressure' SEM, amongst many others, depending upon manufacturer. There is ongoing discussion in the microscopy community as to adopting a generic term to encompass all of these instrument types. For the time being, I propose to use the term variable pressure-environmental SEM (VP-ESEM), with the proviso that this is merely for the sake of convenience.

Our knowledge of the physics of VP-ESEM has only now matured to a level where we can start to collect the concepts together in a dedicated book. There are undoubtedly developments still to come, and the future will bring books that tackle the scientific and technological aspects in much greater depth. What I hope to achieve with this book is a guide that will help those that are just starting out with VP-ESEM, as well as those with more experience looking to gain a deeper appreciation of the concepts.

The principles and applications have been outlined in a generic way, applicable to readers familiar with any of the types of VP-ESEM on the market, irrespective of manufacturer. The aim is to provide a practical overview: the reader is then referred to appropriate sources in the literature should they wish to obtain further information about the inherent physics and chemistry of a particular process or phenomenon. A considerable amount of effort has gone into recognising the work of all those that have contributed to the beginnings, development and growth of this subject. A daunting task in itself. I know that a lot of very interesting and useful work has been carried out, and reported at conferences and meetings, that may not have made it into this book.

One of the biggest difficulties in the field of VP-ESEM is that there is no simple rule that defines which parameters to use. Every specimen and its imaging history will be different and many of the operating parameters are interdependent as well as specimen-dependent. These are the factors that make VP-ESEM so interesting and powerful. With that in mind, an effort has been made to supply a quantitative background to the physics of VP-ESEM, designed to be of help in deciding which set of conditions are appropriate for a given specimen or experiment. I strongly encourage all users of the VP-ESEM to freely experiment for themselves and, using the information contained here and in the literature, to consider the effects of operating conditions on image formation and microanalysis. This truly is the best way to get maximum information from a given specimen.

On a personal note, I would like acknowledge the primary mentors from my early days in this field at the Cavendish Lab, namely Steve Kitching, Brad Thiel and Athene Donald. I am also very grateful to numerous people for their help and advice during the preparation of the manuscript, particularly JJ Rickard, Tony Edwards, Milos Toth, Ralph Knowles, Matthew Phillips, Gerard van Veen, David Joy, Joe Michael and Andrew Bleloch, and to the staff and Executive Committee of the Royal Microscopical Society, especially its current President, Mark Rainforth. A special mention and thanks go to Richard Young for reading the entire manuscript and giving much-needed support and encouragement. Thanks also to those that supplied images and diagrams particularly David Scharf for permission to use the image on the front cover. Last but not least, I'd like to acknowledge the forbearance of my sons Matthew and Oliver.

Debbie Stokes
Cambridge, May 2008

1

A Brief Historical Overview

1.1 SCANNING ELECTRON MICROSCOPY

1.1.1 The Beginnings

The birth of scanning electron microscopy (SEM), in the 1930s and 40s, represented a major breakthrough in the study of the microstructure, composition and properties of bulk materials. SEM combines high-resolution imaging with a large depth of field, thanks to the short wavelengths of electrons and their ability to be focused using electrostatic and electromagnetic lenses. In addition, the strong interaction of electrons with matter produces a wide variety of useful 'signals' that reveal all kinds of secrets about matter at the microscopic and even nanoscopic level.[1]

The earliest demonstration is attributed to Knoll, who obtained the first scanned electron images of the surface of a solid (Knoll, 1935). In 1938, von Ardenne established the underlying principles of SEM, including the formation of the electron probe and its deflection, the positioning of the detector and ways of amplifying the very small signal current (von Ardenne, 1938a, 1938b). Then Zworykin and his team at RCA Research Laboratories built an SEM which had several important original features (Zworykin *et al.*, 1942). The resolution was about 50 nm which, compared to the performance of the already established transmission electron microscope (TEM), was unfortunately not sufficient to convince people of SEM's usefulness at that time.

---

[1] Convention dictates that features measuring less than 100 nm are termed 'nano'.

Principles and Practice of Variable Pressure/Environmental Scanning Electron Microscopy (VP-ESEM)
D. J. Stokes
© 2008 John Wiley & Sons, Ltd

The cause of the SEM was then taken up in 1948 by Oatley at Cambridge University where, over a number of years, he and his research students built five SEMs of increasingly improved performance. The first of these showed how SEM could reveal the three-dimensional nature of surfaces (McMullen, 1952; 1953), and the students that followed made various important contributions to the development of SEM and its applications, leading to an instrument with 10 nm resolution by the 1960s. Some examples of the literature at that time include: Smith and Oatley (1955); Smith (1956); Oatley and Everhart (1957); Wells (1957); Everhart and Thornley (1960); Broers (1965) and Pease and Nixon (1965). The culmination of this work was the production of the first commercially available SEM in 1965: the Cambridge Instruments Stereoscan (see Figure 1.1).

This marked the start of a new era. The resolution of SEM was not as good as that of TEM, but the difficulties of preparing thin samples for TEM were avoided. In addition, it became appreciated that the ability to observe the surfaces of bulk specimens, to visualise the topography of the features and to obtain quantitative information was highly valuable in its own right.

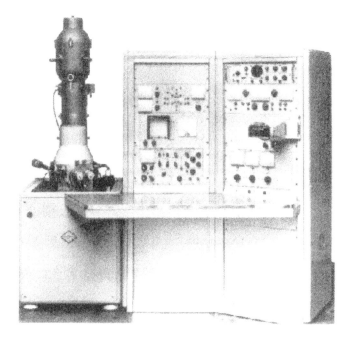

Figure 1.1 The first commercially available scanning electron microscope, the Cambridge Instruments Stereoscan Mk1, 1965

1.1.2 The Need for Added Capabilities

For applications involving metallic materials, SEM imaging and analysis is a comparatively straightforward matter, subject to a proper interpretation of the results and an understanding of the factors that can affect these (such as cleanliness and roughness of surfaces, oxide formation, etc). However, numerous methods are needed when dealing with most other types of material, due to practical operational limits of the instrument and the physics of electron beam–specimen interactions.

To begin with, a fundamental requirement of SEM is the need for high-vacuum conditions throughout the column, typically around $10^{-3}–10^{-5}$ Pa ($10^{-5}–10^{-7}$ torr), sometimes better, depending on the electron source, in order to minimise primary electron scattering and hence maintain a focused beam. An immediate consequence of the high vacuum requirement is that specimens must be vacuum-friendly: no volatile components may be present in the specimen, since this would compromise the vacuum as well as putting the electron source at risk of contamination. Of course, many biological specimens, foams, emulsions, food systems and so on contain water and/or oils – substances that evaporate in the absence of their corresponding vapour. Hence, before imaging can take place, such samples require preparation in order to remove potentially volatile substances, and many procedures have been developed. These include chemical fixing, dehydration in a graded alcohol series, freeze-drying and critical point drying. The methods can be very sophisticated and/or time consuming. An added factor is that the sample preparation technique itself can often change the structural or chemical nature of the specimen to be examined, leading to the imaging of unwanted artefacts.

Moreover, high-vacuum electron microscopy of specimens in the liquid state is, of course, impossible, unless cryogenic procedures are employed to render the specimen solid. It should be emphasised, however, that the methods associated with cryo-preparation are extremely effective for high-resolution observation of frozen-hydrated material and, similarly, the other techniques mentioned certainly have their place. As always, it is a matter of choosing techniques that are appropriate to the system under study.

Another consideration in SEM is that the bombardment of samples by relatively high-energy electrons quickly results in a build up of negative charge unless the sample is electrically conductive, in which case the charge can be dissipated via a grounded specimen holder. Thus, metallic samples, being electrically conductive and containing no volatile components, can be imaged with ease in SEM. Less conductive samples dissipate negative charge much less efficiently and therefore charge builds up. The electric fields in and around the sample quickly become distorted, leading to a deterioration in image quality, often so serious that the sample cannot be imaged at all. Figure 1.2 illustrates the well-known 'mirror effect'.

In the case shown in Figure 1.2, an insulating specimen has first been imaged using a primary electron beam energy E_0 equal to 20 keV, followed by imaging at $E_0 = 3$ keV. For $E_0 = 20$ keV, an excess of electrons is implanted, setting up a strong negative potential below the surface. The 3 keV primary electrons, being much lower in energy, are influenced by the negative potential inside the specimen to such an extent that they are turned back in the opposite direction without entering the solid, striking the polepiece and other fixtures in the chamber and generating electron signals that are collected at the detector. Hence, the specimen surface acts as a mirror.

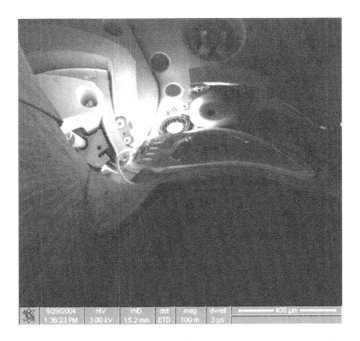

Figure 1.2   The 'mirror effect'. Instead of landing on the specimen and forming an image of the specimen surface, primary electrons are repelled by the electric field arising from electrons implanted in the specimen, and turn back to strike the lens and other parts of the microscope. This generates signals that form an image of the inside of the chamber (distorted in this case)

Assuming that the specimen is not already charged, a low-energy electron beam (arbitrarily a few tens to a couple of thousand electronvolts, eV) can be used so that the number of electrons emitted from the specimen is equal to the number of incident electrons,[2] thus maintaining a charge balance. However, this can be at the expense of image resolution[3] and it can be difficult to find the right criteria for charge balance when the specimen consists of materials with differing electrical properties.

The story so far, then, is that insulators and, very often, the types of samples that have undergone the preparatory stages mentioned earlier, must be subjected to further treatment in the form of a metallic coating. Commonly, insulating samples are sputter-coated with a conductive material such as gold, platinum, palladium, chromium, etc. Again, the introduction of artefacts is a possibility, along with the risk of obscuring fine structural details under the coating. Coated samples give only topographic contrast, due to the short escape depths of electrons from metals, and therefore valuable compositional contrast from the underlying specimen may be lost.

Another consequence arising from the imaging criteria discussed above is that it can be difficult to carry out dynamic experiments, such as mechanical testing, on insulating samples. Even if the sample is given a conductive coating, fracturing of the surface will expose fresh insulating material and lead to charge build up. That said, there are examples of successful results obtained with high-vacuum SEM using low beam energies and/or a backscattered electron (BSE) detector. Electrons forming the latter signal have relatively high energies and are therefore less sensitive to the electric fields that develop as a result of charge build up, compared to low-energy signal carriers such as secondary electrons.

More convincingly, direct, real-time SEM observations of reactions involving gases or liquids are clearly not possible in high vacuum. Such studies are conventionally carried out by observing separate samples, suitably prepared (i.e. fixed, dried, frozen, coated, etc.) at each different stage in the development of the process under study. Clearly, it would be useful if observations could be carried out dynamically, *in situ*, without the preparation steps and vacuum constraints.[4]

This brief introduction has outlined a few of the constraints that conventional high-vacuum SEM places on accessing information from

---

[2] Electrons arriving from the primary electron beam.

[3] This is primarily because low-energy electrons are more easily affected by imperfections in the electron optics, leading to a less tightly focused beam. In modern SEM, though, low-energy resolution is much improved.

[4] The capability to do just this is one of the unique differentiators of VP-ESEM, enabling time-resolved dynamic observations on a single specimen.

certain types of specimen or for performing specific experiments. The desire to go beyond the imaging of extensively prepared, static specimens opened up the way for the development of a new type of SEM, as we shall see in the next section.

1.2 THE DEVELOPMENT OF IMAGING IN A GAS ENVIRONMENT

1.2.1 Overcoming the Limits of Conventional SEM

Since about the 1950s, workers have been experimenting with differentially pumped, aperture-limited 'environmental chambers' for TEM, while others have worked with sealed containers with thin film, electron-transparent windows (see, for example, Swift and Brown, 1970; Parsons, 1975). Then, in 1970, Lane demonstrated the use of an aperture-limited chamber for SEM, described in a relatively obscure, but detailed, conference paper (Lane, 1970). Lane discusses the design of an 'environmental control stage' as well as the scattering cross-sections and mean free paths of electrons in various gases, including hydrogen, oxygen, nitrogen and the noble gases, and demonstrates stable imaging of liquid water.

Meanwhile, Robinson, Moncrieff and others in the 1970s worked to develop an SEM that was capable of maintaining a relatively high pressure while affording controlled imaging, by adapting the SEM specimen chamber itself (Robinson, 1975). Stable imaging of water was also shown by Robinson at the International Congress on Electron Microscopy (ICEM 8) in Canberra, Australia (Robinson, 1974). He was working with a modified JEOL JSM 2 SEM, containing a $100\,\mu m$ pressure-limiting aperture (PLA) to separate the vacuum at the electron source from the specimen chamber at higher pressure. The maximum water vapour pressure was 665 Pa (5 torr), and liquid water was maintained by cooling the chamber and surrounding the specimen with an ice/water reservoir. A solid-state backscattered electron (BSE) detector was used, with reasonable resolution up to a magnification of 2000x. The presence of the aperture restricted the scan range of the electron beam such that the field of view was limited and the minimum magnification was 100x (Robinson, 1996).

At that time, the goal was to enable biological specimens to be imaged without specimen preparation. It was incidentally observed that imaging uncoated insulators at pressures above $\sim$10 Pa seemed to reduce the effects of charging. Early explanations for this centred on the proposition that a film of liquid water was responsible for conferring the necessary conductivity. Of course, this could not explain why imaging with gases other than water had a similar effect. Moncrieff *et al.* (1978) then proposed that it was the collisions between electronic species and gas molecules, resulting in the production of positive ions (see von Engel, 1965), and the attraction of these ions to the negatively charged specimen, which was the mechanism for the observed charge 'neutralisation'.

Moncrieff and co-workers went on to calculate the effects and amount of scattering of primary electrons in, for example, nitrogen gas (Moncrieff *et al.*, 1979). An important conclusion of this work was that, although some primary electrons may be scattered tens to thousands of microns away from their original trajectories, the electrons forming the focused probe maintain a beam of the same diameter as would be formed in high vacuum. This is a vitally important and often misunderstood concept. Meanwhile, the scattered electrons reduce the total current in the focused probe while adding a uniform component to the overall background signal.

In 1978 Robinson began to commercialise this new technology with his company ETP Semra Pty Ltd, manufacturing a device called an environmental cell modification and later called the charge-free anti-contamination system (CFAS). They were mostly sold in Japan, via Akashi/ISI SEMs, and had a pressure limit of 266 Pa (2 torr). In 1980, Akashi/ISI integrated the CFAS and launched WET SEM.

Ultimately, Robinson and co-workers were aiming to work at physiological pressures and temperatures. Observation of liquid water at body temperature ($\sim 37\,^\circ$C) requires a vapour pressure of water above 6.65 kPa (50 torr) and a gas path length[5] no greater than 0.5 mm. Since it becomes physically difficult or impractical to work with a hydrated specimen any closer than this to the pressure-limiting aperture, these criteria define the upper pressure limit. Now, in order to maintain the pressure differential between just two zones separated by one aperture, for a chamber pressure of 6.65 kPa, the aperture size would need to be reduced to $\sim 13\,\mu$m (Robinson, 1996). This places a very large restriction on the field of view. Hence, to improve the situation, Danilatos and Robinson (1979) introduced a two-aperture system, having three differentially pumped zones so that the aperture did not have to be reduced in size. At the same time, Shah and Beckett (1979) were obtaining similar

---

[5] The distance a primary electron has to travel through a gas, defined as the distance between the final pressure-limiting aperture and the surface of the specimen.

results in the UK, using the acronym MEATSEM (moist environment ambient-temperature SEM), and Neal and Mills (1980) built such a system using a Cambridge Stereoscan Mk II.

The next important milestones occurred in the 1980s when Danilatos, working with Electroscan, developed environmental SEM (ESEM). ESEM had a pressure limit of 2.66 kPa (20 torr)[6] and incorporated a secondary electron (SE) detector that could be used in a gaseous environment, utilising the ionisation cascade of secondary electron signals (Danilatos, 1990b). These two features represented a huge increase in the range of commercially available imaging capabilities.

In the 1990s, several other instruments became available in a growing number of geographical areas. These included the low-vacuum SEM, (LVSEM, JEOL); natural SEM (NSEM, Hitachi); environment-controlled SEM (ECO SEM, Amray Inc); and the EnVac (Gresham Camscan). In 1995 the high-tech companies Leica and Carl Zeiss pooled their electron microscopy resources in an independent joint venture, LEO Electron Microscopy Ltd, and introduced variable pressure SEM (VPSEM), later adding an extended pressure (EP) mode. Variable/extended pressure SEM then became products incorporated into Carl Zeiss SMT in 2001. Meanwhile, in 1996, controlled pressure SEM, CPSEM, was introduced by Philips Electron Optics. During that same year, Philips acquired Electroscan and its ESEM technology group, and subsequently merged with FEI, which became known simply as FEI Company in 2002.

Thus, imaging in a gaseous environment became a global phenomenon and now occupies a large fraction of the SEM market. However, something that is clear from the above discussion is that the marketing strategies of the various companies involved have resulted in myriad names and endless confusion. There is no standard, generic term, and this makes it very difficult to talk about the technology without using a specific brand or trademarked name. However, two terms that have emerged as the most appealing are 'variable pressure' and 'environmental'. These terms tend to be used in different contexts: variable pressure suggesting use of a gas for charge control; environmental suggesting that there is an additional need for some specific gas, pressure and/or temperature. Often, the distinction between the two becomes clear from the context of the experiment. Ideally then, it would be more practical to create a single, descriptive term, rather than having to explicitly draw a distinction between techniques that essentially use the same basic

---

[6] The noncommercial version was actually capable of working up to a pressure of 6.65 kPa (50 torr) with a 300 μm aperture.

technology. Hence, in order to simplify the use of acronyms, the generic terminology to be used in this book will be VP-ESEM.

It should be noted that there are several different approaches to signal detection in VP-ESEM, often manufacturer-dependent, for historical reasons. Several of these are summarised below, along with a small selection of literature describing the theory and practice. Irrespective of detection mode, the gas environment is central to all of these methods.

- Detection of the ion signal via the induced specimen current (Danilatos, 1989a; Farley and Shah, 1991; Mohan *et al.*, 1998)
- Collection of the gas-amplified secondary electron signal (Danilatos, 1990a; Thiel *et al.*, 1997; Toth *et al.*, 2006)
- Gas luminescence in which photons generated in the gas are collected and photo-multiplied (Danilatos, 1989b; Morgan and Phillips, 2006)
- Use of a conventional Everhart–Thornley secondary electron detector, physically separated from the higher pressure specimen chamber (Jacka *et al.*, 2003; Slowko, 2006)

1.2.2 Leaps and Bounds

From the earlier work discussed in Section 1.2.1, a number of refinements to the theory and practice of VP-ESEM were set to follow, as the range and quality of experiments began to increase.

The processes underlying gas cascade signal amplification were investigated in more detail (Meredith *et al.*, 1996; Thiel *et al.*, 1997) and the properties of different imaging gases were being explored (Fletcher *et al.*, 1997; Fletcher, 1997; Stowe and Robinson, 1998). Calculations were being made and experimental measurements performed to assess the primary electron beam profile, taking into account electrons that are deflected to form the beam 'skirt' (Mathieu, 1999; Wight *et al.*, 1997; Gillen *et al.*, 1998; Phillips *et al.*, 1999; Thiel *et al.*, 2000; Wight and Zeissler, 2000; Tang and Joy, 2005). As a corollary to this, work was going on in earnest to determine to what extent the presence of a chamber gas influences the results of X-ray microanalysis, particularly for quantitative work. Theories were postulated and algorithms were written to correct for the spurious X-rays generated by scattered primary electrons (Griffin and Nockolds, 1996; Bilde-Sorensen and Appel, 1997; Doehne, 1997a; Gauvin, 1999; Mansfield, 2000; Newbury, 2002; Le Berre *et al.*, 2007).

Meanwhile, several experiments had already begun to suggest that there may be some novel contrast mechanisms available in VP-ESEM, like those found in high-vacuum SEM (such as voltage contrast) but with the added complication that the effects were dependent on specific VP-ESEM operating conditions (Clausen and Bilde-Sorensen, 1992; Horsewell and Clausen, 1994; Harker *et al.*, 1994; Meredith and Donald, 1996). From about 1997 onwards, there was a great deal of activity in the VP-ESEM community following the presentation of findings by Griffin (1997) and Doehne (1998). Whilst observing highly polished mineral specimens, they observed that particular features were only visible under certain conditions, attributed to a charge-related mechanism (Baroni *et al.*, 2000). At around the same time, work was being carried out to explain the contrast observed in heterogeneous liquids (Stokes *et al.*, 1998; Stokes, 1999), and a similarly transient effect was noticed in these systems, too (Stokes *et al.*, 2000).

The potential for VP-ESEM in the study of life science specimens was also being further explored and assessed, and methodologies were being developed that enabled delicate, hydrated biological materials to be observed without laborious preparation (Farley *et al.*, 1988, 1990; Gilpin and Sigee, 1995; Stokes, 2001).

This growing body of emerging observations prompted a series of 'roadmap' meetings in Australia (1999 and 2001 – Figure 1.3) and the USA (2005), aimed at gathering some of the most active members of the

Figure 1.3   Part of a roadmap meeting in the Blue Mountains, Sydney, Australia, 2001. Left to right: Dominique Drouin, Debbie Stokes, Ralph Knowles and Milos Toth. Photograph courtesy of Matthew Phillips, University of Technology, Sydney

community together, in the interests of understanding and advancing this new technology and its applications. And, as the pieces of the jigsaw began to fall into place, it was recognised that something was missing: other than being useful for charge control, what effects were the positive ions generated in the gas cascade process having on the system as a whole? Extensive investigations began that added vital extra detail and further helped to explain some of the charge-related phenomena previously seen (Craven *et al.*, 2002; Toth *et al.*, 2002a; Toth *et al.*, 2002b).

Various aspects of VP-ESEM fundamentals and applications can be found in a number of review articles, for example Newbury (2002); Donald (2003); Stokes (2003b); Muscariello *et al.* (2005); Thiel and Toth (2005), in a special issue of *Microscopy and Microanalysis* (Multi-authors, 2004) and in a few book chapters (Doehne, 1997b; Baker *et al.*, 2003; Stokes, 2003a; Donald, 2007; Griffin, 2007).

It is notable that the annual number of peer-reviewed journal publications has grown steadily over the past decade or so (see Figure 1.4). Given the wide range of names and acronyms in use, numerous variations have been input as search terms in compiling this data. In addition, a wealth of conference papers has been presented in this field. However, conference abstracts are not reflected in the graph.[7]

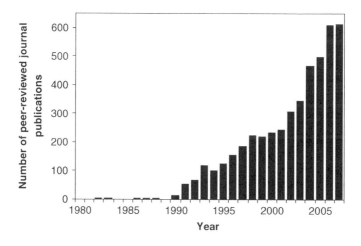

Figure 1.4 Graph to show the number of peer-reviewed journal publications on the subject of 'VP-ESEM'. Source: ISI Web of Knowledge

---

[7] Indeed, it should be noted that reference to conference abstracts has been kept to a minimum throughout this book, given the difficulty in obtaining such material and the often brief, speculative nature of the work reported.

It is gratifying indeed to see that a significant range of applications has now been reported. Many of these will be outlined in Chapter 6. Despite a great deal of early scepticism from the SEM community, ironically similar to the misgivings of the TEM community when the idea of SEM first came along, VP-ESEM has become established as a tremendously versatile technique and the instrument has firmly taken its place as a member of the electron microscopy family.

REFERENCES

Baker, F.S., Craven, J.P. and Donald, A.M. (2003). The environmental scanning electron microscope and its applications, in *Techniques for Polymer Organisation and Morphological Characterisation*. R.A. Pethrick and C. Viney, John Wiley & Sons, Ltd, 111–143.

Baroni, T.C., Griffin, B.J., Browne, J.R. and Lincoln, F.J. (2000). Correlation Between Charge Contrast Imaging and the Distribution of Some Trace Level Impurities in Gibbsite. *Microsc. Microanal.*, 6, 49–58.

Bilde-Sorensen, J. and Appel, C.C. (1997). *X-ray spectrometry in ESEM and LVSEM: corrections for beam skirt effects*. SCANDEM-97.

Broers, A.N. (1965). *Selective ion beam etching in the scanning electron microscope*, University of Cambridge, PhD thesis.

Clausen, C. and Bilde-Sorensen, J. (1992). Observation of voltage contrast at grain boundaries in YSZ. *Micron Microscop. Acta*, 23(1/2), 157–158.

Craven, J.P., Baker, F.S., Thiel, B.L. and Donald, A.M. (2002). Consequences of Positive Ions upon Imaging in Low vacuum SEM. *J. Microsc.*, 205(1), 96–105.

Danilatos, G. (1989a). Mechanisms of Detection and Imaging in the ESEM. *J. Microsc.*, 160(pt. 1), 9–19.

Danilatos, G.D. (1989b). U.S. Patent No. 4,992,662: *Multipurpose gaseous detector device for electron microscope*, ElectroScan Corporation.

Danilatos, G.D. (1990a). Theory of the Gaseous Detector Device in the Environmental Scanning Electron Microscope. *Adv. Electron. Electron Phys.*, 78.

Danilatos, G.D. (1990b). U.S. Patent No. 4,897,545: *Electron detector for use in a gaseous environment*, ElectroScan Corporation.

Danilatos, G.D. and Robinson, V.N.E. (1979). Principles of scanning electron microscopy at high specimen chamber pressures. *Scanning*, 2, 72–82.

Doehne, E. (1997a). A New Correction Method for High-Resolution Energy-Dispersive X-Ray Analyses in the Environmental Scanning Electron Microscope. *Scanning*, 19(2), 75–78.

Doehne, E. (1997b). ESEM development and application in cultural heritage conservation, in *In-Situ microscopy in Materials Research*. P.L. Gai, Kluwer Academic Publishers, pp. 47–65.

Doehne, E. (1998). Charge Contrast: Some ESEM Observations of a New/Old Phenomenon. *Microsc. Microanal.*, 4(Suppl.2: Proceedings), 292–293.

Donald, A.M. (2003). The use of environmental scanning electron microscopy for imaging wet and insulating materials. *Nature Mat.*, 2(8), 511–516.

Donald, A.M. (2007). Environmental scanning electron microscopy, in *Tissue Engineering Using Ceramics and Polymers*, J. Gough, Woodhead.

Everhart, T.E. and Thornley, R.F.M. (1960). Wide-band detector for micro-microampere low-energy electron currents. *J. Sci. Inst.*, 37, 246–248.

Farley, A.N., Beckett, A. and Shah, J.S. (1988). *Meatsem (Moist Environment Ambient-Temperature Scanning Electron-Microscopy) Or The Sem Of Plant Materials Without Water-Loss.* Institute Of Physics Conference Series (93): 111–112.

Farley, A.N., Beckett, A. and Shah, J.S. (1990). *Comparison of Beam Damage of Hydrated Biological Specimens in High-Pressure Scanning Electron Microscopy and Low-Temperature Scanning Electron Microscopy.* Proc. XIIth International Congress for Electron Microscopy, San Francisco Press.

Farley, A.N. and Shah, J.S. (1991). High-Pressure Scanning Electron-Microscopy Of Insulating Materials – A New Approach. *J. Microsc.–Oxford*, 164, 107–126.

Fletcher, A.L. (1997). *Cryogenic Developments and Signal Amplification in Environmental Scanning Electron Microscopy.* University of Cambridge, PhD thesis.

Fletcher, A., Thiel, B. and Donald, A. (1997). Amplification measurements of Potential Imaging Gases in Environmental SEM. *J. Phys. D: Appl. Phys.*, 30, 2249–2257.

Gauvin, R. (1999). Some theoretical considerations on X-ray microanalysis in the environmental or variable pressure scanning electron microscope. *Scanning*, 21(6), 388–393.

Gillen, G., Wight, S., Bright, D. and Herne, T. (1998). Quantitative secondary ion mass spectrometry imaging of self-assembled monolayer films for electron beam dose mapping in the environmental scanning electron microscope. *Scanning*, 20, 400–409.

Gilpin, C.J. and Sigee, D.C. (1995). X-ray microanalysis of wet biological specimens in the environmental scanning electron microscope. 1. Reduction of specimen distance under different atmospheric conditions, *J. Microsc.*, 179(1), 22–28.

Griffin, B.J. (1997). A New Mechanism for the Imaging of Non-conductive Materials: An Application of Charge-Induced Contrast in the Environmental Scanning Electron Microscope (ESEM). *Microsc. Microanal.*, 3(Suppl. 2: Proceedings), 1197–1198.

Griffin, B.J. (2007). Variable Pressure and Environmental Scanning Electron Microscopy. Imaging of Biological Samples, in *Electron Microscopy Methods and Protocols*, second edition. J. Kuo, Humana Press.

Griffin, B.J. and Nockolds, C.E. (1996). *Quantitative EDS analysis in the ESEM using a bremsstrahlung intensity-based correction for primary electron beam variation and scatter.* Microscopy and Microanalysis '96.

Harker, A.B., Howitt, D.G., Denatale, J.F. and Flintoff, J.F. (1994). Charge-Sensitive Secondary-Electron Imaging Of Diamond Microstructures. *Scanning*, 16(2), 87–90.

Horsewell, A. and Clausen, C. (1994). *Voltage contrast of ceramics in the environmental SEM.* ICEM 13, Paris, France.

Jacka, M., Zadrazil, M. and Lopour, F. (2003). A differentially pumped secondary electron detector for low-vacuum scanning electron microscopy. *Scanning*, 25, 243–246.

Knoll, M. (1935). Aufladepotentiel und Sekundäremission elektronenbestrahlter Körper. *Z. tech. Phys.*, **16**, 467–475.

Lane, W.C. (1970). *The environmental cold stage*. Proceedings of the Third Scanning Electron Microscopy Symposium, IIT Research Institute, Chicago, IL, 60616.

Le Berre, J.F., Demers, H., Demopoulos, G.P. and Gauvin, R. (2007). Skirting: A limitation for the performance of X-ray microanalysis in the variable pressure or environmental scanning electron microscope. *Scanning*, **29**(3), 114–122.

Mansfield, J.F. (2000). X-ray microanalysis in the environmental SEM: A challenge or a contradiction? *Microchim. Acta*, **132**(2–4), 137–143.

Mathieu, C. (1999). The beam–gas and signal–gas interactions in the variable pressure scanning electron microscope. *Scanning Microsc.*, **13**(1), 23–41.

McMullen, D. (1952). *Investigations relating to the design of electron microscopes*, University of Cambridge, PhD thesis.

McMullen, D. (1953). An improved scanning electron microscope for opaque specimens. *Proc. Inst. Electr. Engrs.* **100**(pt II), 245–259.

Meredith, P. and Donald, A.M. (1996). Study of 'wet' polymer latex systems in environmental scanning electron microscopy: some imaging considerations. *J. Microsc.*, **181**(pt.1), 23–35.

Meredith, P., Donald, A.M. and Thiel, B. (1996). Electron–gas interactions in the environmental scanning electron microscope's gaseous detector. *Scanning*, **18**(7), 467–473.

Mohan, A., Khanna, N., Hwu, J. and Joy, D.C. (1998). Secondary electron imaging in the variable pressure scanning electron microscope. *Scanning*, **20**, 436–441.

Moncrieff, D.A., Barker, P.R. and Robinson, V.N.E. (1979). Electron scattering by the gas in the scanning electron microscope, *J. Phys. D: Appl. Phys.* **12**(4), 481–488.

Moncrieff, D.A., Robinson, V.N.E. and Harris, L.B. (1978). Gas neutralisation of insulating surfaces in the SEM by gas ionisation. *J. Phys.D: Appl. Phys.*, **11**(17), 2315–2325.

Morgan, S.W. and Phillips, M.R. (2006). Gaseous scintillation detection and amplification in variable pressure scanning electron microscopy. *J. Appl. Phys.*, **100**(7), Article no. 074910.

Multi-authors (2004). Special issue: Characterisation of Nonconductive Materials. *Microsc. Microanal.*, **10**(6).

Muscariello, L., Rosso, F., Marino, G., Giordano, A., Barbarisi, M., Cafiero, G. and Barbarisi, A. (2005). A critical review of ESEM applications in the biological field. *J. Cell. Phys.*, **205**, 328–334.

Neal, R.J. and Mills, A. (1980). Dynamic Hydration Studies In An SEM. *Scanning*, **3**(4), 292–300.

Newbury, D.E. (2002). X-ray microanalysis in the variable pressure (environmental) scanning electron microscope. *J. Nat. Inst. Standards Technol.*, **107**(6), 567–603.

Oatley, C.W. and Everhart, T.E. (1957). The examination of p-n junctions in the scanning electron microscope. *J. Electronics*, **2**, 568–570.

Parsons, D. (1975). Radiation Damage in Biological Materials, in *Physical Aspects of Electron Microscopy and Microbeam Analysis*. B. Siegel, New York: pp. 259–265.

Pease, R.F.W. and Nixon, W.C. (1965). High resolution scanning electron microscopy. *J. Sci. Inst.*, **42**, 81–85.

Phillips, M.R., Toth, M. and Drouin, D. (1999). A Novel Technique for Probe Intensity Profile Characterization In the Environmental Scanning Electron Microscope. *Microsc. Microanal.*, 5(2), 294–295.

Robinson, V.N.E. (1974). *A wet stage modification to a scanning electron microscope.* Proc. 8th Int. Cong. Electron Microscopy, Canberra, Aust. Acad. Sci.

Robinson, V.N.E. (1975). The elimination of charging artifacts in the scanning electron microscope. *J. Phys. E: Sci. Instr.* 8(8), 638–640.

Robinson, V.N.E. (1996). SEM at high chamber pressures. http://www.microscopy.com/MicroscopyListserver/MicroscopyArchives.html (November 19, 1996).

Shah, J.S. and Beckett, A. (1979). A preliminary evaluation of moist environment ambient temperature scanning electron microscopy (MEATSEM). *Micron*, 10, 13–23.

Slowko, W. (2006). New system for secondary electron detection in variable-pressure scanning electron microscopy. *J. Microsc.–Oxford*, 224, 97–99.

Smith, K.C.A. (1956). *The scanning electron microscope and its fields of application.* University of Cambridge, PhD thesis.

Smith, K.C.A. and Oatley, C.W. (1955). The scanning electron microscope and its fields of application. *Br. J. Appl. Phys.*, 6, 391–399.

Stokes, D.J. (1999). *Environmental SEM studies of the microstructure and properties of food systems*, University of Cambridge, PhD thesis.

Stokes, D.J. (2001). Characterisation of Soft Condensed Matter and Delicate Specimens using Environmental Scanning Electron Microscopy (ESEM). *Adv. Eng. Mat.*, 3(3), 126–130.

Stokes, D.J. (2003a). Investigating Biological Ultrastructure using Environmental Scanning Electron Microscopy (ESEM), in *Science, Technology and Education of Microscopy: An Overview. No. 1, Vol. II.* A. Mendez-Vilas. Badajoz, Spain, Trans Tech Publications Ltd, pp. 564–570.

Stokes, D.J. (2003b). Recent advances in electron imaging, image interpretation and applications: environmental scanning electron microscopy. *Philosoph. Trans. Roy. Soc. Lond. Series A – Math, Phys. Eng. Sci.*, 361(1813), 2771–2787.

Stokes, D.J., Thiel, B.L. and Donald, A.M. (1998). Using Secondary Electron Contrast for Imaging Water–Oil Emulsions in the Environmental SEM (ESEM). *Microsc. Microanal.*, 4(Suppl. 2: Proceedings), 300–301.

Stokes, D.J., Thiel, B.L. and Donald, A.M. (2000). Dynamic Secondary Electron Contrast Effects in Liquid Systems Studied by Environmental SEM (ESEM). *Scanning*, 22(6), 357–365.

Stowe, S.J. and Robinson, V.N.E. (1998). The use of helium gas to reduce beam scattering in high vapour pressure scanning electron microscopy applications. *Scanning*, 20, 57–60.

Swift, J.A. and Brown, A.C. (1970). Environmental Cell for Examination of Wet Biological Specimens At Atmospheric Pressure By Transmission Scanning Electron Microscopy. *J. Phys. E – Sci. Instr.*, 3(11), 924.

Tang, X.H. and Joy, D.C. (2005). An experimental model of beam broadening in the variable pressure scanning electron microscope. *Scanning*, 27(6), 293–297.

Thiel, B.L., Bache, I.C., Fletcher, A.L., Meredith, P. and Donald, A.M. (1997). An Improved Model for Gaseous Amplification in the Environmental SEM. *J. Microsc.*, 187(pt. 3), 143–157.

Thiel, B.L., Bache, I.C. and Smith, P. (2000). Imaging the probe skirt the environmental SEM. *Microsc. Microanal.*, **6**(Suppl. 2), 794–795.

Thiel, B.L. and Toth, M. (2005). Secondary electron contrast in low-vacuum/environmental scanning electron microscopy of dielectrics. *J. Appl. Phys.*, **97**(5).

Toth, M., Daniels, D.R., Thiel, B.L. and Donald, A.M. (2002a). Quantification of electron–ion recombination in an electron-beam-irradiated gas capacitor. *J. Phys. D – Appl. Phys.*, **35**(14), 1796–1804.

Toth, M., Knowles, W.R. and Thiel, B.L. (2006). Secondary electron imaging of nonconductors with nanometer resolution. *Appl. Phys. Lett.*, **88**(2), Article no. 023105.

Toth, M., Phillips, M.R., Thiel, B.L. and Donald, A.M. (2002b). Electron Imaging of Dielectrics under Simultaneous Electron–Ion Irradiation. *J. Appl. Phys.*, **91**(7), 4479–4491.

von Ardenne, M. (1938a). Das Elektronen-Rastermikroskop. Praktische Ausführung. *Z. tech. Phys.*, **19**, 407–416.

von Ardenne, M. (1938b). Das Elektronen-Rastermikroskop. Theoretische Grundlagen. *Z. Phys.*, **109**, 553–572.

von Engel, A. (1965). *Ionized Gases*, Clarendon Press, Oxford.

Wells, O.C. (1957). *The construction of a scanning electron microscope and its application to the study of fibres*, University of Cambridge, PhD thesis.

Wight, S., Gillen, G. and Herne, T. (1997). Development of environmental scanning electron microscopy electron beam profile imaging with self-assembled monolayers and secondary ion mass spectroscopy. *Scanning*, **19**, 71–74.

Wight, S.A. and Zeissler, C.J. (2000). Direct measurement of electron beam scattering in the environmental scanning electron microscope using phosphor imaging plates, *Scanning* **22**(3), 167–172.

Zworykin, V.A., Hillier, J. and Snyder, R.L. (1942). A scanning electron microscope. *ASTM Bull*, **117**, 15–23.

2

Principles of SEM

2.1 INTRODUCTION

Many of the basic principles of imaging apply to both conventional high-vacuum SEM and VP-ESEM. This chapter is therefore dedicated purely to the operation of high-vacuum SEM, so that we have a firm basis on which to develop the concepts of VP-ESEM later. In any case, the VP-ESEM also doubles as an excellent high-vacuum instrument, and so knowledge of each mode will be valuable. Although only relatively short descriptions are given here, much more detailed information about SEM can be found in textbooks like, for example, Reimer (1985); Newbury *et al.* (1986); Goodhew *et al.* (2001) and Goldstein *et al.* (2003).

2.1.1 Why Use an Electron Beam?

Electrons can be extracted from various sources and driven by an electrical potential along an evacuated column. Electrons generated in this way are called primary electrons, and they can be formed into a finely focused beam and systematically scanned across a surface of interest. When primary electrons strike a specimen surface, a wide range of useful interactions can occur, causing various charged particles and photons to be generated. Those that are emitted can be collected and used to form an image, diffraction pattern or chemical spectrum. In addition, for thin specimens, primary electrons can be transmitted through the material and similarly utilised. Typical signals are shown schematically in Figure 2.1.

Principles and Practice of Variable Pressure/Environmental Scanning Electron Microscopy (VP-ESEM)
D. J. Stokes
© 2008 John Wiley & Sons, Ltd

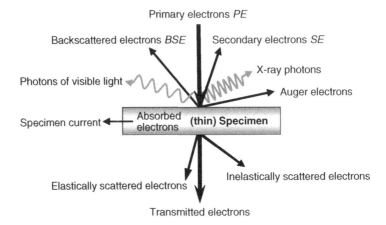

Figure 2.1 Some of the useful signals that are generated when a focused electron beam strikes a specimen. Note that, for a 'thick' specimen, i.e. more than a few hundred nanometres, electrons become absorbed within the specimen and hence are not transmitted

In SEM, the principal electron signals that are used are backscattered electrons (BSEs) and secondary electrons (SEs). Differences in specimen composition and surface topography affect the generation, transport and escape of these signals. Images formed in an SEM result from variations in electron signal intensity collected at each point (pixel) as the electron beam briefly dwells within the scanned area. We will come back to these interactions in detail in Section 2.4.

In addition to the strong interactions of electrons with matter, their mass and charge make it easy to direct them to where we want them, using electromagnetic fields to act as lenses and mirrors, rather like glass lenses and deflectors are used in a light microscope.

Crucially, the size of features that can be resolved is primarily determined by the wavelength λ of the probing radiation: the shorter the wavelength, the smaller the feature that can be seen. For comparison, a few figures-of-merit are shown below.

Light: $\lambda = 5 \times 10^{-7}$ m
X-rays: $\lambda = 1 \times 10^{-10}$ m
Electrons: $\lambda = 1 \times 10^{-11}$ m (for accelerating voltages $\sim$30 kV)[1]
            $\lambda = 1 \times 10^{-12}$ m (for accelerating voltages around 100 kV)

---

[1] A simplified expression for electron wavelength is given by $\lambda = (1.5/V_0)^{1/2}$, in nanometres. V_0 is the accelerating voltage: the potential difference that accelerates electrons from the source down the column. For the voltages used in SEM (up to 30 kV), this expression is fine. For higher voltages, as used in TEM, relativistic effects would need to be taken into account.

So, the shorter wavelength of electrons gives a distinct increase in achievable resolution compared to light. In fact, the resolution for the most advanced TEMs is better than 0.1 nm, enabling individual atoms to be seen. For modern SEM, the maximum resolution is better than 1 nm. This is mainly limited by the physics of electron–specimen interactions. More on this later (Section 2.4).

Aside from the technological aspects, one of the main differences between TEM and SEM is the specimen thickness. In TEM, specimens must be just a few tens to hundreds of nanometres thick, whereas SEM specimens can be several centimetres thick and many centimetres across, depending on the size of the specimen chamber. This flexibility of sample handling often gives SEM an advantage compared to TEM, in terms of ease and time of sample preparation, for situations where the ultimate resolution capabilities of TEM are not a necessary requirement.

An advantage of SEM compared to the light microscope is the larger depth of field, that is to say, how far above and below the actual plane of focus details can still be resolved clearly. We will return to this in Section 2.5.7.

2.1.2 The SEM Column

A simplified schematic diagram of an SEM system is depicted in Figure 2.2. The system basically consists of:

- an electron source;
- lenses and apertures;
- coils for rastering (scanning) the beam;
- control electronics and high-voltage supplies;
- a deflector/acquisition system for collecting and processing the signal information;
- a monitor to display the information;
- a vacuum system for the source, column and specimen chamber.

2.1.3 Why Do We Need a Vacuum System?

The SEM electron source (see Section 2.2) generally operates under a vacuum of 10^{-3} to 10^{-5} Pa (10^{-5}–10^{-7} torr), depending on the type of electron source[2] and the method of pumping.

---

[2] Some sources (i.e. cold field emission) require ultra high vacuum conditions on the order of 10^{-8} Pa (10^{-10} torr).

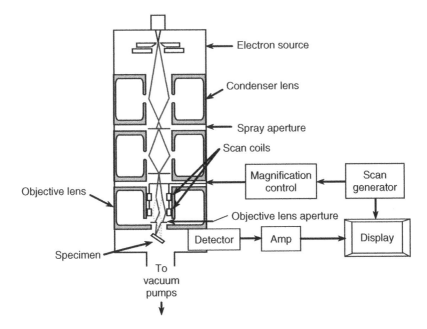

Figure 2.2 Simplified schematic diagram of the basic components of an SEM. Adapted from Goldstein *et al.* (2003)

There are two good reasons for this need for a vacuum system:

- the electron source is easily contaminated;
- electrons scatter off anything that gets in the way.

Traditionally, in addition to the region around the electron source, the rest of the electron column is held at a similarly high vacuum, to help maintain the unscattered trajectories of primary electrons on their way to the specimen. However, in VP-ESEM, the pressure is deliberately increased in the region of the specimen. The reasons for and implications of this will be discussed in detail in Chapters 3 and 4.

2.2 ELECTRON SOURCES

To produce a beam of electrons, a high voltage is applied to a filament.[3] The filament may be either a thermionic emitter or a field emission source, and these will be discussed briefly in Sections 2.2.1 and 2.2.2,

---

[3] Also known as an electron gun, although this term is gradually being phased out.

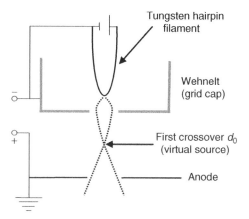

Figure 2.3 Simplified schematic diagram of a thermionic tungsten hairpin emission source

respectively. The emitted electrons are controlled by an electrode that focuses them into a crossover of diameter d_0 (see Figure 2.3) and convergence semi-angle α_0 (as shown in Figure 2.8).

The brightness of the source β is a very important parameter, and is given by the current density (amps A per m^2) per unit solid angle (in steradians) (written as $A\cdot m^{-2} \, sr^{-1}$ or $A/m^2 \, sr$), and is given by:

$$\beta = j_b/(\pi\alpha^2) = 4i_b/(\pi^2 d_0 \alpha^2) \tag{2.1}$$

where j_b is the beam current density, i_b is the emission current, $\alpha = \alpha_0$ is the convergence semi-angle and d_0 the crossover diameter of the beam. The brightness of the source increases linearly with accelerating voltage V_0.

2.2.1 Thermionic Emission Sources

Typically, thermionic filaments are made either of tungsten in the form of a v-shaped 'hairpin' or a lanthanum hexaboride (LaB_6) single crystal, formed into a pyramidal point. These filament types are resistively heated to release electrons (hence the term thermionic) as they overcome the minimum energy needed to escape the material (the work function Φ).[4]

Figure 2.3 is a simplified diagram to show the principle of electron beam formation using a tungsten hairpin filament. A negative bias placed

---

[4] For tungsten, the work function $\Phi = 4.5\,eV$ and for LaB_6, $\Phi \sim 2.5\,eV$.

on the Wehnelt cylinder (also known as the grid cap), together with an anode at ground potential, forces the beam into the first crossover d_0. This is essentially an image of the emission area of the filament, and is referred to as the virtual source.

2.2.2 Field Emission Sources

A LaB_6 filament gives a more coherent beam and its brightness (5×10^9–10^{10} A·m^{-2} sr^{-1}) is much higher than the tungsten hair-pin (5×10^8 A·m^{-2} sr^{-1}). Electrons are emitted from a smaller area of the LaB_6 filament, giving a source size of about 1 µm, compared to around 50 µm for the tungsten hairpin. This leads to greatly improved image quality with the LaB_6 source. In addition, the lifetime of a LaB_6 source is considerably longer than for a tungsten hairpin (roughly 1000 hours vs 100 hours), although a better vacuum is required for the LaB_6, 10^{-5} Pa (10^{-7} torr), compared with 10^{-3} Pa (10^{-5} torr) for tungsten.

Field emission sources[5] are much brighter (10^{13} A·m^{-2} sr^{-1}) and more stable than thermionic emitters. They exhibit a much narrower energy spread, making a larger fraction of the emitted electrons useful (see the discussion on chromatic aberration in Section 2.3.2.2).

For a field emission source, a fine, sharp, single crystal tungsten tip is employed. Very high extraction fields are applied ($>10^9$ V/m), by placing a high negative potential on the tip (see Figure 2.4), allowing electrons to tunnel through the surface work function barrier.

The two anodes focus the primary beam to a crossover, and the second anode accelerates (or frequently decelerates) the electrons to the final desired energy. Because the electron extraction process is independent of the final beam voltage, field emission sources are ideal for low-voltage imaging in SEM.

For a so-called cold emission source, heating of the filament is not required (i.e. it operates at room temperature). However, this type of filament is prone to contamination and requires more stringent vacuum conditions (10^{-8} Pa, 10^{-10} torr). Regular and rapid heating ('flashing') is required in order to remove contamination. The spread of electron energies is very small for a cold field emitter (0.3 eV) and the source size is around 5 nm.

Other field emission sources, known as thermal and Schottky sources, operate with lower field strengths. The thermal source, as the name

---

[5] For historical reasons, the field emission source and associated electrodes are also collectively known as the field emission gun, and frequently abbreviated to FEG.

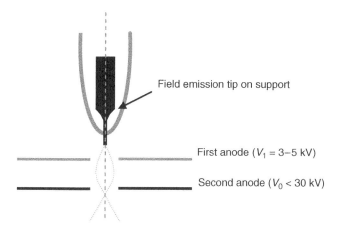

Field emission tip on support

First anode (V_1 = 3–5 kV)

Second anode (V_0 < 30 kV)

Figure 2.4 Schematic diagram of a field emission source. The first anode controls the potential applied to the tip (V_1), while the second anode provides the accelerating voltage (V_0)

suggests, works in conjunction with an elevated temperature to lower the barrier for electron tunnelling. Here, the source size is also 5 nm, but the energy spread is 1 eV or so. The Schottky source is also heated, and dispenses zirconium dioxide onto the tungsten tip to further lower its work function. The Schottky source is actually a thermionic emitter, but has similar characteristics to the cold and thermal field emission sources, in that it has the same high brightness and a small energy spread (comparable to the thermal emitter). The source size is slightly larger, 20–30 nm.

2.3 ELECTRON OPTICS

2.3.1 Lenses

A series of electromagnetic lenses is used to shape and focus the electron beam. Electromagnetic lenses consist of many thousands of windings of copper wire inside a soft iron shell (the polepiece). An example is shown in Figure 2.5.

When an electric current is passed through the windings, both electric E and magnetic B fields are generated, applying a force F to the electrons in the beam according to the Lorentz equation:

$$F = e(\vec{E} + \vec{v} \otimes \vec{B}) \tag{2.2}$$

where v is the velocity of the electrons in the beam.

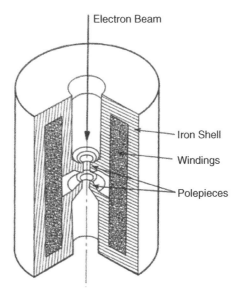

Figure 2.5 Diagram of an electromagnetic lens. The yoke in the centre confines the field to a very small region. Reproduced from Goldstein *et al.* (2003), with permission from Springer (Plenum Kluwer)

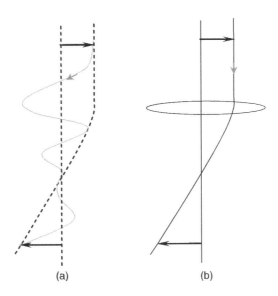

Figure 2.6 (a) Diagram showing how electrons passing through an electromagnetic lens are brought to focus; (b) a conventional optical ray diagram for comparison. Electrons rotate along this path (rotation omitted for clarity). Notice how the envelope marked by the dashed line in (a) resembles that of the ray in (b)

As electrons pass through the lens, they feel both radial and circumferential forces and so begin to spiral towards the centre of the lens (the optical axis), bringing the beam to focus. This is shown diagrammatically in Figure 2.6(a). Notice that the 'envelope' around the path of the electron beam resembles the ray path that would normally depict a light ray (see Figure 2.6(b)).

Increasing the current in the lens increases the field strength and causes the crossover point of the beam to move upwards. Likewise, reducing the current moves the crossover downwards.

The lens system of an SEM is summarised in Figure 2.7. Note that the convention here is to represent lenses in the same way as for light optics (i.e. glass lenses) and to draw the corresponding ray traces. Although the action of focusing in the electron microscope is guided by electromagnetic fields rather than glass lenses, this is a convenient way to visualise the behaviour of the primary electrons (by analogy with Figure 2.6).

The electrons first pass through the condenser lens (sometimes two, usually coupled together and operated as one) which reduces the size

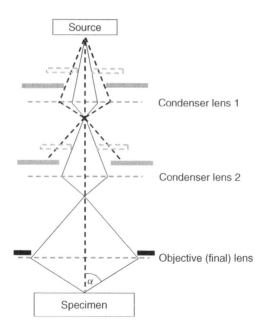

Figure 2.7 The lens system in an SEM. Variable apertures (light grey) in the condenser lens system influence the spread of electrons and control the beam current, while the objective lens aperture (dark grey) determines the convergence semi-angle α

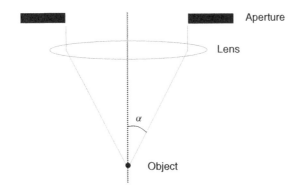

Figure 2.8 Definition of the convergence semi-angle α subtended by the objective aperture

(demagnifies) the crossover d_0 formed at the electron source. The condenser lens plays an important role in determining the final beam size (and, hence, resolution) as well as the beam current.

Increasing the current in the condenser lens increases the convergence semi-angle α and reduces the beam spot size. Meanwhile, an aperture in the condenser lens limits the angular range of electrons that are allowed to travel on to the objective lens, where a final aperture may be positioned. This determines the beam's final convergence semi-angle α, as Figure 2.8 demonstrates. Note that decreasing the aperture size restricts the available beam current.

Adjusting the focus control changes the current in the objective lens, moving the crossover point up and down according to the height of the specimen. Hence, the focal plane of the lens can be made to coincide with the plane of the object to form a focused image. The distance between the end of the lens polepiece and the specimen is known as the working distance, *WD*.

2.3.2 Lens Aberrations

Electromagnetic lenses tend to introduce aberrations that affect the focusing of the electron beam. These aberrations are the limiting factor in determining the ultimate resolution of an instrument. In various ways, they cause electrons to be focused in slightly different planes, and result in the beam having a finite minimum diameter, rather than being an infinitely sharp point. This minimum diameter is called the disc of least confusion.

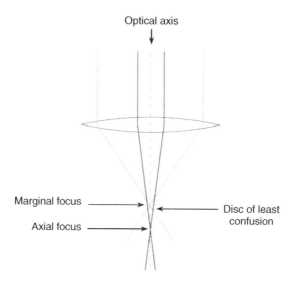

Figure 2.9 Spherical aberration C_s causes electrons arriving at different distances from the optic axis of the lens to be focused in different planes

2.3.2.1 Spherical Aberration

A very important lens defect is called spherical aberration. Electrons that arrive at the lens far from the optic axis are focused more strongly than electrons entering the lens closer to the axis. This is shown in Figure 2.9.

Lens quality is often given by the coefficient of spherical aberration C_s. The minimum beam diameter d_s is a function of C_s and the convergence semi-angle α:

$$d_s = 1/2C_s\alpha^3 \tag{2.3}$$

Note that C_s is proportional to the focal length f of the lens and so lenses with short f (for example, snorkel or immersion lenses as opposed to conical lenses[6]) are preferable for minimising the effects of spherical aberration and maintaining a tightly focused primary beam. In addition, the aperture contained in the condenser lens helps to stop electrons that would otherwise travel towards the edge of the objective lens, and a further aperture could be employed in the objective lens itself (although recall that the use of apertures restricts beam current and, as we will see in Section 2.3.2.3, can cause diffraction aberration).

---

[6] For a description of these lens types, the reader is referred to Goldstein *et al.* (2003), for example.

2.3.2.2 Chromatic Aberration

Another important type of aberration is chromatic aberration, particularly at low beam energies. This time, electrons are focused in different planes due to the spread in electron energies ΔE arriving at the lens (Figure 2.10). The smaller the energy spread of the primary beam electrons, the smaller the effect of the coefficient of chromatic aberration C_c. In this case, the minimum beam diameter d_c is a function of C_c, convergence semi-angle α and the fractional variation of electron energies $\Delta E/E_0$, where E_0 is the primary beam energy:

$$d_c = (\Delta E / E_0) C_c \alpha \qquad (2.4)$$

This is where field emission sources, with a smaller energy spread ΔE, have a big advantage over thermionic sources (as mentioned in Section 2.2.2, field emitters have ΔE of roughly $0.3-0.7$ eV, compared to about 1.5 eV for lanthanum hexaboride and 3 eV for a tungsten hairpin). The effects of C_c can also be reduced by increasing the primary beam energy E_0 and by using a smaller aperture. However, there are limits to how small the aperture can be, not just because of limits on beam current, but also because of diffraction effects, as discussed below.

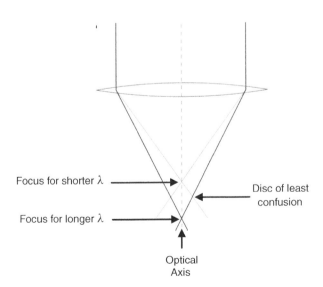

Focus for shorter λ

Disc of least confusion

Focus for longer λ

Optical Axis

Figure 2.10 Ray diagram to show the effects of chromatic aberration C_c. Electrons with different energies (and therefore wavelengths λ) arriving at the same point in the lens are focused to different planes

2.3.2.3 Aperture Diffraction

When electrons pass through a small aperture, they are diffracted such that, rather than producing a single spot at the image plane, they form a series of circular rings (an Airy disc intensity distribution). This means that the image of a point has a Gaussian distribution about a central maximum, as shown in Figure 2.11. Given the reciprocal nature of diffraction, the smaller the aperture, the larger will be the central spot.

The increase in diameter of the beam as a function of aperture diffraction d_d is given by:

$$d_d = 0.61\,\lambda/\alpha \qquad\qquad (2.5)$$

According to Equation (2.5), as the convergence semi-angle α increases, the effect of aperture diffraction becomes smaller. Conversely, spherical aberration and, to a lesser extent, chromatic aberration, increase with increasing angle. Hence, there is an optimum angle that satisfies the compromise between these situations for a given wavelength (or energy) of electrons.

2.3.2.4 Astigmatism

The final type of defect that concerns us here is astigmatism, caused by asymmetries in the lenses or dirty apertures. Effectively, this means that there are two different focal lengths in perpendicular directions, and

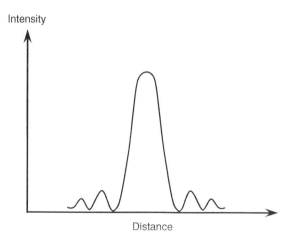

Figure 2.11   Schematic plot to show the effects of aperture diffraction

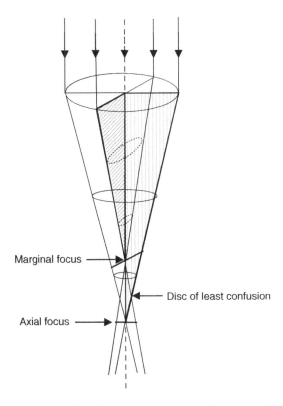

Marginal focus

Disc of least confusion

Axial focus

Figure 2.12 Diagram to show how astigmatism produces an enlarged beam spot size due to differences in focus in the horizontal and vertical planes. Astigmatism is easily corrected by applying an additional, weak magnetic field to the lens

again this causes the overall beam diameter to be larger than it should be. This is shown schematically in Figure 2.12.

Astigmatism can be seen by changing the focus above and below the focal plane. A smearing of the image will be observed which changes direction $(x-y)$. This can be corrected using stigmators which apply a counter field in the x- and y-directions to reshape the beam to a circular cross-section in these planes and, in so doing, reduce the beam diameter and improve resolution.

When working at magnifications above about 10 000x, it is essential to regularly check and correct for astigmatism, and get into a routine (focus, stigmate, focus), to achieve the best results possible. Figures 2.13 and 2.14 show some example images to demonstrate the effect.

Ultimately then, the final beam (or probe) diameter d_p is determined by a combination of source size, condenser lens strength, aperture size(s) and various aberrations.

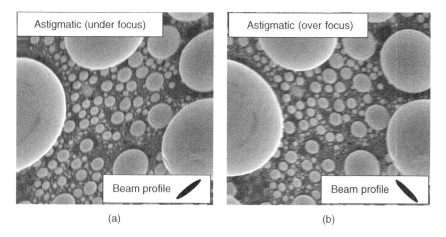

Figure 2.13 The effects of astigmatism. Under- and over-focused images, (a) and (b) respectively, show that the beam is distorted in opposite directions.

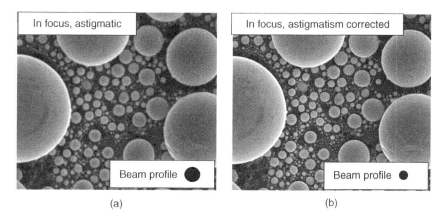

Figure 2.14 The effect of astigmatism on focus. In (a), the image has been focused without correcting for astigmatism: the beam diameter is therefore not optimal and the image appears slightly blurred. After applying x- and y-stigmators, astigmatism is corrected; (b) the beam diameter is reduced and the image is in sharper focus. Images courtesy of Ellen Baken, FEI Company

2.4 SIGNALS AND DETECTION

A primary electron (PE) makes random elastic and inelastic collisions until either it loses all of its energy and comes to rest within the sample, also called thermalisation, or it reaches a boundary of the sample and is emitted as a backscattered electron (BSE) or as a transmitted

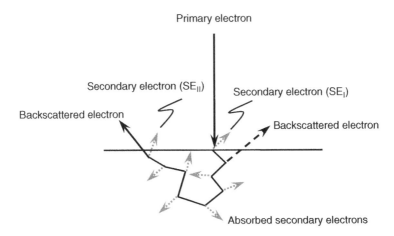

Figure 2.15 Simplified schematic diagram to show the path of a primary electron and the emission of backscattered electrons and secondary electrons (types SE_I and SE_{II})

electron (if using a thin sample). As previously shown in Figure 2.1, other signals generated by a primary electron beam include secondary electrons (SEs), Auger electrons, characteristic X-rays and photons. Each signal originates from a specific volume within the sample, potentially yielding sample characteristics such as chemical composition, surface topography, phase distributions and crystallinity. The simplified diagram in Figure 2.15 shows schematically the path of a primary electron and the emission of backscattered electrons (BSEs) and secondary electrons (SEs).

We will discuss electron emission in detail in Sections 2.4.2 and 2.4.3, but note that two types of secondary electron are shown in Figure 2.15: those generated by the primary electron beam (termed SE_I) and those caused by backscattered electrons (termed SE_{II}).

Meanwhile, Figure 2.16 schematically illustrates the typical range 'R' (i.e. length) and spatial distributions within a region of sample (the interaction volume) characterising the various signals that can be produced by an electron beam. The ranges and escape depths of signals vary according to the type and energy of the interaction. In particular, signals such as secondary electrons and Auger electrons come from only a very tiny portion of the total interaction volume, since they lack the energy to travel large distances.

The size and shape of the interaction volume varies as a function of both the energy of the primary beam and the nature of the sample. It is

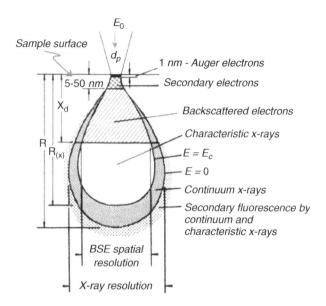

Figure 2.16 Schematic diagram to show the interaction volume of the primary electron beam. Adapted from Goldstein *et al.* (1992)

worth exploring the main ways that a primary electron can interact with matter in a little more detail, in order to understand the nature of some of the various electrons and photons produced and the information they convey. Some of the many texts on the subject of electron interactions with solids include those of Dekker, 1958; Jenkins and Trodden, 1965; Bishop, 1974; Seiler, 1983; Reimer, 1985; Kaneko, 1990; Bongeler *et al.*, 1992; Joy, 1995; Goodhew *et al.*, 2001; Goldstein *et al.*, 2003.

2.4.1 Primary Electrons and the Interaction Volume

2.4.1.1 Elastic Interactions

Elastic interactions affect the trajectories of the primary beam electrons inside the specimen without altering their kinetic energies. An elastic collision occurs when a primary electron is within range of the electrostatic field of an atomic nucleus. The electron is scattered due to the Coulombic attraction of the nucleus, partially screened by orbital electrons, causing the electron to deviate from its path by an angle ϕ_e. This angle can be anything between 0° and 180° (0–π radians), although it is typically 2°–5° per scattering event.

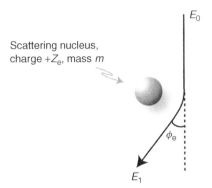

Figure 2.17   Schematic diagram to show the elastic (Rutherford) scattering of an electron by the Coulombic attraction of a nucleus. E_0 represents the energy of the incoming primary electron and E_1 that of the deflected electron, where $E_0 \approx E_1$. Adapted from Goldstein *et al.* (2003)

Figure 2.17 illustrates the elastic scattering of an incident electron in which the direction component of the electron vector is changed, but the magnitude of the velocity v remains almost unchanged, so that the electron's kinetic energy $(1/2m_e v^2)$ is also unchanged (where m_e is the mass of the electron).

The probability that a primary electron will interact with an atom is known as its cross-section σ, which varies according to the energy of the electron. The elastic cross-section σ_e is inversely proportional to E_0^2, the square of the primary beam energy, and proportional to Z^2, the square of the atomic number, so that the probability of elastic scattering is greater for low-energy primary electrons traversing a material of high atomic number. This is summed up in the expression for the (Rutherford) elastic scattering cross-section σ_e, given below:

$$\sigma_e(> \phi_0) = 1.62 \times 10^{-20}(Z^2/E^2)\cot^2(\Phi_0/2) \qquad (2.6)$$

where ϕ_0 is a given scattering angle, in radians, and $\sigma_e\ (> \phi_0)$ denotes the probability of scattering an electron to angles larger than ϕ_0. Note that $E = E_0$, in keV.

The units of Equation (2.6) are: events $(>\phi_0)$/electron·(atom/cm^2) or, more simply, σ_e is expressed in terms of the cross-sectional area, in cm^2, to describe the probability of an elastic scattering event. The dependence on Z arises from the fact that the higher the atomic number, the greater the number of protons in the nucleus and so the stronger the Coulombic attraction.

Using Equation (2.6) we can also predict the distance that an electron will travel before encountering an elastic collision: its elastic mean free path λ_e. This can be determined from:

$$\lambda_e = A/N_0\rho\sigma_e \qquad (2.7)$$

where A = atomic weight (g/mole), N_0 = Avogadro's number, ρ = density (g/cm^3) and σ_e is calculated from Equation (2.6).

In addition, Monte Carlo simulations can be used to predict the range of primary electrons as they interact with a sample, and there are several easy-to-use computer programs available for this purpose. An excellent example is Wincasino (Hovington *et al.*, 1997; Drouin *et al.*, 2007. See also http://www.gel.usherbrooke.ca/casino/).

2.4.1.2 Inelastic Interactions

Inelastic interactions result in the transfer of energy from primary electrons to the atoms of the sample. In the context of electron microscopy, this energy exchange is limited to atomic electrons, rather than nuclei, since nuclei are very difficult to excite. Atomic electrons, being quantised, accommodate extra energy by moving to a higher orbital (the process of excitation) or by leaving the atom altogether (ionisation). Inelastic scattering decreases the kinetic energy of the bombarding electron, whilst the deviation of its path is very small, on the order of 0.1° or less.

Inelastic scattering leads to the production of secondary electrons, Auger electrons, characteristic and Bremsstrahlung (continuum) X-rays, electron-hole pairs in insulators and semiconductors, long-wavelength electromagnetic radiation in the visible, ultraviolet and infrared regions of the electromagnetic spectrum (cathodoluminescence), lattice vibrations (phonons) and collective electron oscillations (plasmons). Rather than using individual scattering cross-sections for these processes, it is convenient to group them together into an average rate of (continuous) energy loss dE/ds (known as stopping power), where s is the distance travelled in the sample. An expression by Bethe (1930) approximates dE/ds (in terms of keV/cm) for primary beam energies above ~5 keV:

$$\frac{dE}{ds} = -7.85 \times 10^4 \frac{z\rho}{Ae_m} \ln\frac{1.116E_m}{J} \qquad (2.8)$$

where ρ = density of the sample (g/cm^3), E_m = average energy along the path of the electron (keV), J = mean ionisation potential (eV) (sometimes approximated as $0.115Z$ eV) and A = molecular weight.

A modified form of this equation has been proposed (Joy and Luo, 1989) to take account of the effect of decreasing contributions of inelastic scattering processes at beam energies below 5 keV.

2.4.1.3 Dependence of Interaction Volume on Primary Electron Energy

A low-energy primary electron is more likely to undergo a collision with an atom of the sample, since it moves more slowly than a high-energy electron. In addition, low-energy electrons lose energy at a faster rate than high-energy electrons. The distances between collisions (mean free path λ) and penetration depth of primary electrons into the sample are therefore smaller when the beam energy is low, and so the interaction volume is smaller at low beam energies compared to high.

For example, Figure 2.18 shows a Monte Carlo simulation comparing the behaviour of primary electrons in silicon for beam energies $E_0 = 5$ keV and 30 keV. Although the two data sets appear very similar, a closer inspection of the scales reveals that primary electrons travel to a depth (y-axis) and width (x-axis) more than 20 times farther at the higher energy.

2.4.1.4 Dependence of Interaction Volume on Atomic Number

Primary electrons penetrating a material with a low atomic number have a longer mean free path λ than electrons with the same energy in a material with a high atomic number. This is due to consideration of the Coulombic attraction of the nucleus, where a higher atomic number leads to a greater Coulombic attraction (see Section 2.4.1.1).

So, electrons of a given energy undergo a greater degree of (elastic) scattering per unit length when the atomic number is high, reducing the size of the interaction volume. In addition, the scattering angles of primary electrons in high atomic number materials are larger. This is demonstrated in Figure 2.19, where the lower Z material is represented by carbon and the higher Z material is represented by gold.

2.4.1.5 Beam Penetration

As we have seen in the previous section, primary electrons do not travel far in metals and so the interaction volume is very small relative to, for example, carbon, in which electrons can travel much larger distances. This generally holds for all materials with low atomic numbers, and

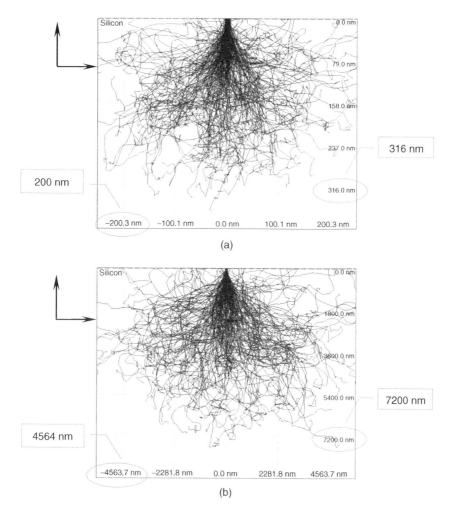

Figure 2.18 The interaction volumes for low vs high primary electron energy E_0 in a silicon substrate ($Z = 14$). (a) $E_0 = 5\,\text{keV}$; (b) $E_0 = 30\,\text{keV}$. The lateral spread and penetration depth are much increased in the case of (b). See text for discussion. Simulated using CASINO v.2.42 (Drouin *et al.*, 2007)

also has important implications for the escape of signal electrons, as we shall see in the next sections. Further, the penetration and escape of electrons is of great significance in VP-ESEM, where specimens are generally imaged without a conductive (metallic) coating.

A consequence of these variations in electron range becomes apparent when we have a mixture of materials in one specimen, in which case the penetration depth of the primary electrons will also vary in different

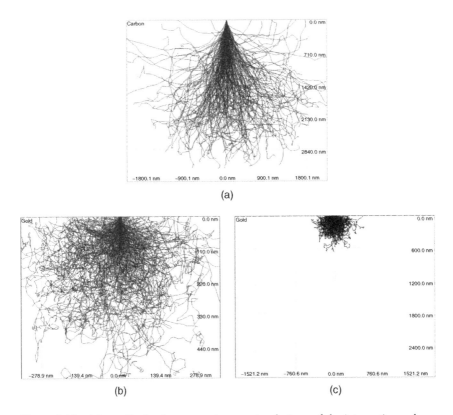

Figure 2.19 Monte Carlo electron trajectory simulations of the interaction volume at 20 keV and 0° tilt for elements of different atomic number Z: (a) carbon (low Z) and (b) gold (higher Z). Note that the scaling in each case is quite different. The range of electrons in carbon is roughly six times larger than for gold. (c) The range of 20 keV electrons in gold using a similar scale to that in (a). Also notice that the shape of the interaction volume changes from a 'teardrop' shape to roughly spherical. Simulated using CASINO v.2.42 (Drouin *et al.*, 2007)

regions of the specimen. An example of this can be seen in Figure 2.20, where a copper grid is covered by a thin polymer film (~160 nm thick). Being organic, we can assume that the constituents of the polymer have small atomic numbers (between $Z = 6$ and $Z = 8$, for example), while for copper, $Z = 29$. For a primary beam energy $E_0 = 3$ keV, we see a uniform layer of polymer (except for the folds in the right-hand corner that act as a reference point). As the energy is increased, the beam begins to penetrate the film, generating a signal from the copper grid underneath. At higher energy still, the beam penetrates right through the polymer film, rendering it invisible, while the copper grid is clearly visible.

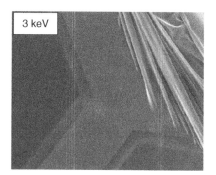 

Figure 2.20 Demonstration of electron beam penetration. A copper grid is coated with a thin polymer film and imaged using the secondary electron signal. As the beam energy is increased, the copper grid becomes increasingly visible and the polymer film less visible. In fact at higher energies, such as $E_0 = 30$ keV shown here, the film is essentially invisible

It is very important, therefore, that before imaging a specimen, the beam energy (or the equivalent accelerating voltage V_0) is set at a sensible value for the specimen under consideration. High voltages on low Z materials will result in large penetration depths and give a false impression of the surface. In addition, the larger penetration depth and increased interaction volume will result in poorer spatial resolution due to the lateral spread of primary and emitted electrons.

2.4.2 Backscattered Electrons

Backscattered electrons (BSEs) are primary beam electrons that have been deflected by collisions with atoms to such an extent that their path actually takes them back up through the sample surface. They can be emitted from deep within the sample. Depending on collision history, BSE energies can range from the primary beam energy E_0 down towards the level of secondary electron energies. However, there are two distinct regions in the energy distribution of backscattered electrons.

Region I is attributed to electrons that have retained $\geq 50\%$ of the primary beam energy. Materials of intermediate and high atomic number Z will have a high proportion of backscattered electrons in this region. Region II is a broad tail of energy distributions, representing primary electrons that have travelled progressively farther into the sample, losing energy (inelastically) before being backscattered (elastically) out of the sample. The lower energy limit associated with backscattered electrons is arbitrarily taken to be 50 eV. These regions are shown in Figure 2.21.

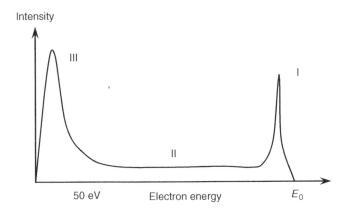

Figure 2.21 Idealised schematic plot to illustrate the distribution of electron signals, and their approximate energies, for a given primary beam energy. Regions I and II refer to the backscattered electron contributions, while Region III corresponds to the secondary electron signal

The BSE coefficient η is the ratio of the number of backscattered electrons emitted n_{BSE} to the number of primary electrons bombarding the sample n_{PE} (or their equivalent currents i):

$$\eta = \frac{n_{BSE}}{n_{PE}} = \frac{i_{BSE}}{i_{PE}} \qquad (2.9)$$

Monte Carlo simulations are a simple way to estimate a value for η.

The BSE coefficient is found to increase smoothly and monotonically with atomic number Z, for a given angle of incidence (apart from deviations when the atomic number is between 22 and 29, attributed to variations in the ratio of atomic weight to atomic number). This dependence on Z provides an important contrast mechanism in SEM, enabling different materials to be distinguished.

Figure 2.22 summarises the behaviour of η as a function of Z, according to the following expression:

$$\eta = -0.0254 + 0.016Z - 1.86 \times 10^{-4}Z^2 + 8.3 \times 10^{-7}Z^3 \qquad (2.10)$$

The BSE coefficient is also found to vary with the tilt angle θ of the sample, and the reason for this originates from the geometry of the escape depth relative to the incident primary beam. This behaviour serves as an important contrast mechanism, in that different topographical features can again be distinguished on the basis of variations in η (a flat surface

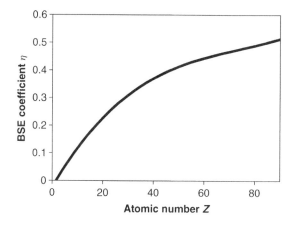

Figure 2.22 Backscattered electron coefficient η as a function of atomic number Z

having a lower coefficient than a sloped surface or edge). A similar effect will be discussed for secondary electrons in Section 2.4.3 (Figure 2.24).

The value of η is relatively insensitive to primary beam energy (above $\sim$5 keV). This is due to two competing factors. As the beam energy E_0 is increased, the probability of elastic scattering of the primary beam is reduced. One might therefore expect the value of η to be correspondingly lower with increasing beam energy. However, scattered electrons with high initial energy lose energy at a lower rate (per unit length of material) than scattered electrons with lower initial energy. The effect is that higher energy backscattered electrons are less readily absorbed by the sample than lower energy backscattered electrons. Hence, the net result is that η remains roughly the same at a variety of beam energies.

2.4.3 Secondary Electrons

Secondary electron excitations result when loosely bound valence electrons are promoted from the valence band to the conduction band in insulators and semiconductors, or directly from the conduction band in metals. These excited electrons then propagate through the sample, experiencing inelastic collisions and energy loss themselves.

Once at the sample surface, the electron can be emitted as a secondary electron. Low-energy SEs are arbitrarily defined as having energies <50 eV, with the majority having energies <10 eV. Typically, the energy distribution is that shown in Figure 2.23, where the peak in energy comes in the range 2–5 eV. The general shape of the distribution is the

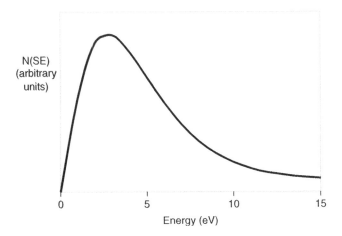

Figure 2.23 A generalised secondary electron emission curve showing the energy distribution with a peak between 2–5 eV

same for all materials, although the maximum value of the emission coefficient and the energy at which this occurs varies.

The SE coefficient δ is defined as the ratio of the number of SEs emitted n_{SE} to the number of primary electrons n_{PE} bombarding the sample (or their equivalent currents i):

$$\delta = \frac{n_{SE}}{n_{PE}} = \frac{i_{SE}}{i_{PE}} \tag{2.11}$$

Unlike BSE, the SE coefficient is relatively insensitive to atomic number Z. A typical value for δ is therefore $\sim$0.1 for most elements, although for carbon, δ takes a value of $\sim$0.05 and for gold δ is $\sim$0.2.

Significantly, because of their low energies, excited electrons travel only short distances, a few nanometres, and so SEs that emerge from the sample must originate close to the surface. The signal from these electrons thus provides information about the surface of the sample and is the principal signal used for topographical information in SEM. In fact, the probability p of SE escape decreases exponentially with depth z, as a result of the strong attenuation of secondary electron energies by inelastic scattering:

$$p \approx \exp \frac{-z}{\lambda} \tag{2.12}$$

where λ = mean free path of excited electrons as they propagate through the sample towards the surface.

The mean free path is equivalent to the mean depth of emission (mean escape depth) of secondary electrons. The value of λ is said to be about 1 nm in metals and up to 20 nm in insulators (although there is actually a range of electron energies, leading to a spread in λ) (Seiler, 1983).

The *maximum* depth of emission T is taken to be about 5λ. Inelastic scattering of SEs by conduction electrons, which are abundant in metals, leads to the reduction in λ for metals, relative to insulators, noted above. The mean escape depth of SEs is around 1/100 that of backscattered electrons. Once at the surface, secondary electrons must have sufficient energy to overcome the work function ϕ of the sample.

As with η, δ has a dependence on the angle of tilt of the sample. This is a consequence of the orientation of the escape depth region of the interaction volume. In addition, surface features such as steps or edges in a sample help to increase the proportion of exposed interaction volume, and thus increase emission. These points are demonstrated in Figure 2.24.

Finally, δ is dependent on primary beam energy to some extent. At high beam energies (10–30 keV, for example) δ is well below unity, however, as the beam energy is lowered to around 5 keV, the total electron yield ($\delta + \eta$) begins to rise. As η is relatively insensitive to beam energy, the increase in yield is attributed to an increase in δ. The point at which the total yield reaches unity is termed the second crossover point E_2. At lower energies still, the yield can actually become greater than unity (i.e. more electrons are emitted from the sample than are supplied). At low beam energies, the total yield falls below unity once again, and

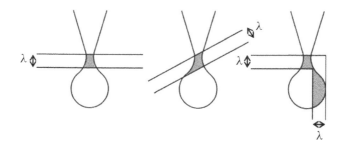

Figure 2.24 Schematic illustration of the tilt angle dependence of secondary electron emission. Only those electrons created within a distance λ of the surface are able to escape (shaded area). As the surface is tilted, a greater proportion of the interaction volume is exposed. Emission at edges is particularly high. (For backscattered electrons, the situation is very similar, although the escape depth region is much greater since backscattered electrons have higher energies)

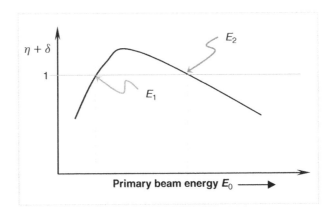

Figure 2.25 Schematic plot of total yield illustrating the increase in yield as the primary beam energy is reduced. Between the crossover points E_1 and E_2 the yield can exceed unity over a range of low primary electron energies, where the generation and escape of secondary electrons is favoured

this is termed the first crossover point E_1 (not to be confused with beam crossover, which is an entirely different matter, see Section 2.2). The total yield behaviour as a function of primary beam energy E_0 is illustrated in Figure 2.25.

The explanation for this behaviour is in the relationship between the penetration of the primary beam and the escape depth of secondary electrons. When the beam energy is below ~5 keV, the interaction volume is small, as primary electrons are more strongly attenuated by scattering. The interaction volume becomes so shallow between E_1 and E_2 that many of the electrons created by the primary beam are within the escape depth of the sample, and thus a greater proportion are able to escape as secondary electrons.

There are a number of sources of secondary electron (SE) signal emanating from the sample and its surroundings in the SEM sample chamber. The most relevant are usually termed SE_I and SE_{II} (Drescher *et al.*, 1970):

- SE_I – originate from the sample surface at the point of impact by the primary electron beam. This is the most important secondary electron imaging signal as it gives the highest resolution.
- SE_{II} – produced by backscattered electrons as they exit the sample surface. For a low Z target, such as carbon, the SE_{II} signal is approximately 1/5 that of the SE_I signal. For a high Z target, such as gold, the SE_{II} signal becomes ~1.5 times larger than the SE_I signal.

The total secondary electron coefficient δ_T is thus a function of both the true, SE_I coefficient δ_{SEI} and the backscattered electron-dependent SE_{II} coefficient $\delta_{SEII}\eta$:

$$\delta_T = \delta_{SEI} + \delta_{SEII}\eta \qquad (2.13)$$

Because η is virtually independent of beam energy, the total number of SE_{II} remains roughly constant with increasing beam energy. However, there is a lateral spread in the interaction volume of BSE. This increases the spatial distribution of SE_{II}, i.e. they may be liberated some distance away from the point of interest.

Other types of SE are formed as BSEs strike the walls of the chamber and primary electrons strike the polepiece. These are sometimes referred to as SE_{III}.

2.4.4 X-ray Radiation

These important species are emitted as a result of the relaxation of an excited state in an atom. The initial excited state is caused by the ionisation of a tightly bound inner electron by an incident primary electron. The process is illustrated in Figure 2.26. Such signals

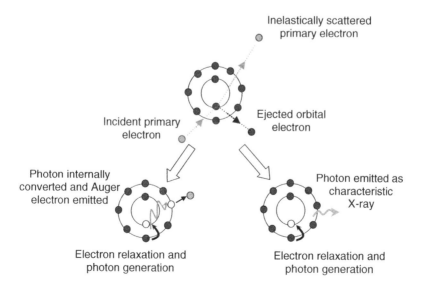

Figure 2.26 Schematic illustration of the process of inner electron ionisation and the production of either a characteristic X-ray or Auger electron

are very useful for deducing the chemical composition of materials, and chemical characterisation in the electron microscope is known as microanalysis, to distinguish it from imaging. The most common form of X-ray microanalysis is called energy dispersive X-ray spectroscopy, abbreviated to EDS or EDX. An alternative form of microanalysis is called wavelength dispersive X-ray spectroscopy, WDS. Auger electron spectroscopy is commonly used in the (scanning) transmission electron microscope (S/TEM), under ultra-high vacuum conditions, and will not be discussed further here.

By exploiting the nature of the primary electron-specimen interactions shown in Figure 2.26, all of these techniques are designed to detect species that carry specific atomic, and therefore chemical, information. X-ray spectroscopy also produces a background signal caused by continuum X-rays. Continuum and characteristic X-rays are discussed briefly in the next two sections. For in-depth discussions, see Reimer (1985), Goldstein *et al.* (2003) and Goodhew *et al.* (2001).

2.4.4.1 Continuum X-rays

X-rays can be generated as a result of the deceleration of primary electrons in the Coulombic field of the nucleus, similar to the process shown in Figure 2.17. Because of the 'braking' effect exerted on the electron, this is also known as Bremsstrahlung radiation. Since the X-ray photons produced can have any energy, up to the energy of the primary beam, these tend to contribute to a background continuum in an X-ray spectrum. A typical background can be seen in Figure 2.27. The background signal can be useful in certain cases, and one example will be discussed in Chapter 5 (Section 5.4.2).

2.4.4.2 Characteristic X-rays

A much more useful type of X-ray emission is that generated by the relaxation of an excited atomic state arising from an inner shell vacancy. The transition of an electron from one of the outer shells produces an X-ray with an energy (or wavelength) that represents the difference between the two states. These energy differences are very specific to the individual atomic species, and the transitions are governed by a set of well-documented selection rules. Figure 2.27 shows a typical EDS spectrum, containing both characteristic X-ray peaks and a background continuum.

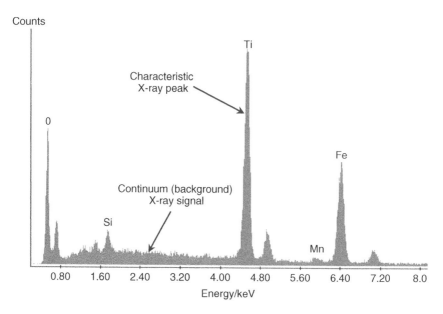

Figure 2.27 A typical spectrum obtained by energy dispersive X-ray spectroscopy, EDS. The peaks carry specific information about the chemistry of atoms in the specimen as a result of inelastic collisions between primary electrons and atoms. The background continuum is caused by energy loss as a primary electron passes close to atoms and feels the Coulombic attraction of the nucleus

2.4.5 Cathodoluminescence

This process is similar to that for characteristic X-ray or Auger electron production, but is related to the relaxation of an excited state in an outer atomic shell. The energy release is generally much smaller and gives rise to the release of a photon of energy in the ultraviolet, visible or infrared region of the electromagnetic spectrum.

Again this is useful, since the electronic band structure of a material determines the energies of cathodoluminescent photons produced. The electronic structures of materials will be outlined in Chapter 5, but a brief explanation is given here. Consider a valence band electron that has been promoted to the conduction band by inelastic scattering of a primary electron. This subsequently leaves a 'hole' in the valence band. If this electron-hole pair recombines, the excess energy is released as a photon of specific energy, equal to that of the energy gap E_g (typically 2–3 electronvolts in semiconductors, up to several electronvolts in insulators). These mechanisms are illustrated in Figure 2.28.

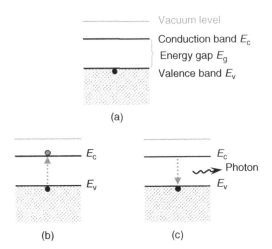

Figure 2.28 Simplified schematic diagram to illustrate the emission of a cathodoluminescent photon from an insulator. (a) The valence band is full and the conduction band is empty; (b) an inelastically scattered primary electron causes the promotion of a valence electron to the conduction band, leading to an electron-hole pair. If the excited electron does not escape the attraction of the hole, then the pair will recombine; (c) emitting a photon with energy E_g as a result

2.5 PRACTICAL ASPECTS OF ELECTRON BEAM IRRADIATION

So, electrons are incredibly useful for creating images of very tiny features and for giving us chemical information. But bombarding a specimen with negatively charged particles can, and frequently does, result in some problems. First, any type of electron–specimen interaction is effectively a form of damage, and so these are outlined in the next section. Secondly, if the specimen is not a particularly good electrical conductor then negative charge accumulates inside the bulk of the specimen and this affects the imaging process. We will explore the various charging phenomena and the benefits of VP-ESEM for these types of samples in more detail in Chapter 5. In the meantime, we will review some of the strategies that can be employed to help reduce charging effects in high-vacuum SEM.

2.5.1 Radiation Damage

A brief description of the causes and effects of radiation damage is presented here. However, for a more thorough treatment of radiation damage mechanisms in electron microscopy, the reader is referred to the following: Glaeser (1975); Isaacson (1975); Parsons (1975); Reimer (1975); Cosslett (1978); Reimer and Schmidt (1984) and Talmon (1987). In particular, the emphasis of this section is on organic materials and water, since these are especially susceptible to the effects of radiation damage, and these processes are the most relevant to the discussions about imaging uncoated materials in VP-ESEM that will follow in later chapters.

The processes leading to radiation damage can be grouped into two main categories, primary and secondary:

Primary:

- excitation of an individual atom or group of atoms (plasmons);
- ionisation of an atom;
- displacement of an atom.

Secondary:

- emission of electrons, X-rays or light;
- temperature increase;
- electrostatic charging;
- bond scission;
- cross-linking;
- mass loss;
- formation of a carbonaceous coating (contamination).

Most radiation damage to organic specimens is caused by plasmon excitations and atomic ionisations. These generally give rise to specimen heating. The rate of radiation-induced chemical transformations can be two to three times higher (or more) when a substance is in a liquid state rather than solid.

Ionisation of water molecules leads to the formation of highly reactive species that can cause damage to the molecules present in a sample. Some reaction schemes for the formation of radiolysis products from water are shown below (after Talmon, 1987), where $\bullet$ represents an unpaired electron (free radical) and * represents an excited state

Primary radiolysis	$H_2O \rightarrow H_2O\bullet^+ + e^-$
Formation of excited water molecule	$H_2O\bullet^+ + e^- \rightarrow H_2O^*$
Return to ground state	$H_2O^* \rightarrow H_2O + heat$
Decomposition into free radicals	$H_2O^* \rightarrow H\bullet + \bullet OH$
Proton transfer	$H_2O\bullet^+ + H_2O \rightarrow H_3O^+ + \bullet OH$
Electron interaction with hydronium ion	$H_3O^+ + e^- \rightarrow H\bullet + H_2O$

In organic materials (polymers and biopolymers included), degradation occurs during electron beam bombardment: scission takes place
at random along the chain and can lead to the production of smaller
molecules. These can diffuse to the surface and leave by sublimation.
Scission is more likely to occur where a molecule has large side groups
(as in proteins, for example). The ends of polymer chains are susceptible
to depolymerisation, which can lead to 'unzipping' of polymers.

Alternatively, cross-linking may occur, following primary ionisation
and the formation of a side chain free radical and a hydrogen free
radical. The H radical diffuses easily to abstract another H radical,
forming H_2, leaving another polymer radical behind. Various lengths
of polymer radicals join up in pairs to form cross-links, leading to a
more radiation-resistant specimen. There is a small amount of mass loss
associated with the process, but this is much less significant than the
eventual loss of crystallinity in the sample. Scission and cross-linking
occur simultaneously, but scission may predominate.[7]

In compounds containing π-bonds, it is thought that excess energy is
emitted as heat or light, rather than leading to excitation or ionisation,
due to the accommodation of energy in the large number of energy
levels present in π-bonds. Such compounds are thus intrinsically relatively radiation-resistant. A substantial fraction of energy absorbed is
dissipated as heat. Normally, beam-heating effects only raise the sample
temperature by a couple of degrees, but if the specimen is incorrectly
mounted or thermal contact to the specimen holder is poor, the temperature can increase by tens of degrees. Diffusion of free radicals and
other reaction products can be reduced at lower specimen temperatures.

In addition, the thickness of the specimen may determine how much
of the electron dose is actually absorbed by the material. For example, if
the specimen is a very thin film (i.e. no more than one or two elastic mean
free paths at the incident energy) and the electron beam has sufficient
velocity, then only a small amount of the energy of the electron beam is
deposited in the sample, and less damage is caused.

2.5.2 Minimising Specimen Charging – Low-Voltage SEM

As we saw in Section 2.4.3, for each material there are primary beam
energies (crossover energies E_1 and E_2 – see Figure 2.25) at which
the number of incoming primary electrons is matched by the number

---

[7] It should be noted that electron beam lithography techniques make use of these beam-induced scission
and cross-linking effects to modify special polymer resists, enabling the creation of high-resolution structures.

of emitted electrons (BSEs and SEs), so that the net charge is zero. Typically, E_2 values for highly insulating materials are in the range 1–3 keV. A description of the upper crossover energy E_2 is given by Joy and Joy (1998), and tabulated values for some materials can be found in Goldstein *et al.*, (2003).

Knowledge of crossover energies means that the primary electron beam can be specifically tuned to avoid any significant charge build up for a given material. This usually involves low beam energies, or voltages, and is generally known as low-voltage SEM, abbreviated to LVSEM.[8]

It can be tricky to satisfy the energy crossover condition, however, when specimens contain a mixture of materials with widely differing electrical properties, since their crossover energies may vary substantially.

Imaging at low voltages brings with it one or two other caveats. Since the interaction volume decreases with beam energy, care must be taken to avoid localised radiation damage, caused by having a high flux of electrons in a small volume in addition to the fact that low-energy electrons interact more strongly with matter. Whilst lateral resolution increases with decreasing interaction volume, because the BSE signal emanates from a smaller region, the electron beam diameter tends to increase at low beam energies, due to increased sensitivity to aberrations and a higher likelihood of electrons interfering with each other,[9] thus reducing the resolution. The confinement of low-energy primary electrons to the near surface region also makes the resulting signals more sensitive to surface contamination.

Nonetheless, low-voltage imaging is clearly a good way to avoid implanting excessive amounts of negative charge in bulk materials, and so the effects of charging can be easier to control. It is also worth pointing out that the backscattered electron signal is less sensitive to charging than the secondary electron signal.[10]

2.5.3 Increasing Surface and Bulk Conductivities

An alternative strategy to using low beam energies is to apply a thin, electrically conductive coating to the surface of the specimen and/or

---

[8] Beware: the acronym LVSEM is also used for low-vacuum SEM.

[9] This is known as the Boersch effect, which increases with increasing beam current density and decreasing beam energy.

[10] Recall that BSEs have higher energies than SEs, and note that high-energy electrons are less affected by electric fields.

to infuse conductive species into the bulk microstructure. There are a number of effects that these measures bring, especially for organic media, including:

- more stable imaging and X-ray analysis over a wider range of beam energies;
- reduction of primary beam penetration/interaction volume, giving:
  - increased secondary electron emission
  - increased surface detail
  - increased lateral resolution;
- increased mechanical and thermal stability;
- improved contrast, especially between materials having low atomic numbers.

2.5.3.1 Electrically Conductive Surface Coatings

Methods for applying conductive coatings include evaporation and sputtering, and there are numerous ways of doing this (Goldstein et al., 2003). Coatings include gold, palladium, platinum, silver, various metal alloys, carbon and, for high-resolution work, iridium, tantalum, tungsten and chromium.

Note that an electrically conductive coating does not make an insulating specimen conductive: it just provides a surface ground plane onto which an electrical field can terminate (see Chapter 5). It is therefore advisable to avoid using high-electron beam voltages or currents, as these could still lead to charge build-up within the specimen.

If applied too thickly, conductive coatings can obscure small features, and a coating can itself introduce artefacts such as surface pitting and cracking.

2.5.3.2 Electrically Conductive Bulk Stains

Solutions or vapours containing conductive heavy metal salts can be infused into bulk organic material or infiltrated into nonconductive porous or fibrous materials. Frequently, such stains involve osmium tetroxide or ruthenium tetroxide, phosphotungstic acid and uranyl acetate (Goldstein et al., 2003).

This type of specimen preparation can take several hours or days, and is often associated with the introduction of artefacts in biological tissues, since organic material is converted into inorganic products.

However, heavy metal impregnation does increase contrast between different features and allows specimens to be imaged without a conductive surface coating.

2.6 SEM IN OPERATION

Now that we have covered the basic principles of primary electron generation, focusing, signal generation and some aspects of how all this affects the specimen, the aim of this concluding section is to draw together some of the other practicalities of imaging in SEM. In particular, there are several factors to take into account before we can truly distinguish between the various features of a specimen, and these are covered throughout the following subsections.

2.6.1 Building Up an Image

The primary electron beam is usually scanned (rastered) over a square area of the specimen. This scanning typically occurs line by line, with each line being scanned from left to right over the area of interest, with a fast fly-back from right to left between lines. The corresponding image is displayed on a CRT or monitor, where the screen raster is synchronous with that of the electron beam. The intensity of each screen pixel is determined by the signal level (intensity) arriving at the detector at each point.

In an 8-bit greyscale image, pixels are assigned one of 256 grey level values (0–255 inclusive), where $0 = $ black and $255 = $ white. Thus, the specimen features become 'visible' by virtue of differences in signal levels at adjacent pixels in the area scanned. Some microscopes offer options to acquire/display 16-bit or 24-bit data per pixel, which support more greyscale levels or colour overlays, respectively.

2.6.2 Magnification

Magnification M in an SEM is determined by the length of the electron beam line scan on the specimen $L_{specimen}$ relative to the size of a line on the display $L_{display}$ onto which the information is relayed, point by point:

$$M = L_{display}/L_{specimen}$$

Figure 2.29 shows how the information is transferred by scanning a given specimen area. This dependence of magnification on the particular display size can lead to confusion when comparing images from

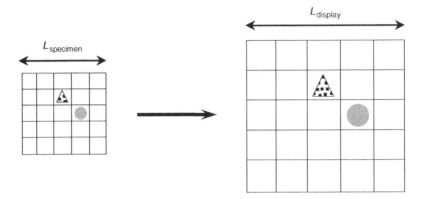

Figure 2.29 The information from the scanned specimen is transferred point by point to a display. Magnification is defined as the ratio of the two line scans and therefore, on its own, magnification is not an absolute measure of specimen feature sizes

different sources since, clearly, magnification is not an absolute measure of an object's size. Hence, it is important to include with an image an independent measurement indicator such as a scale bar or the horizontal field width.

2.6.3 Signal-to-noise Ratio

One of the first requirements in distinguishing different parts of a sample is that there has to be a change in signal level (intensity) ΔS in order to produce contrast (see Figure 2.30(a)). Next, we require an adequate

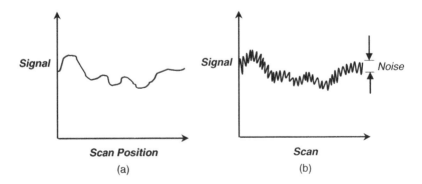

Figure 2.30 As the scan position changes, the signal intensity arising from specimen features may vary, as shown in (a). The signal is modulated by random fluctuations (noise), making the signal more difficult to distinguish

signal-to-noise ratio S/N. Noise is a result of random fluctuations in image intensity, as depicted in Figure 2.30(b).

According to the Rose criterion, an observer requires ΔS to exceed the noise N by at least a factor of five in order to detect a small object against a background:

$$\Delta S > 5N \tag{2.14}$$

The signal-to-noise ratio improves as the mean number of counts n (i.e. the number of discrete events sampled by the detector) increases, since noise becomes a less significant fraction of the total signal.

Signal-to-noise can be expressed as the square root of the mean number of counts:

$$\frac{S}{N} = \frac{n}{n^{1/2}} = n^{1/2} \tag{2.15}$$

So the required signal can be expressed in terms of the number of signal events:

$$\Delta S > 5n^{1/2} \tag{2.16}$$

2.6.4 Contrast

Following on from the discussion on signal-to-noise ratios, we can start to define how much primary electron beam current is needed in order to produce sufficient contrast (after Oatley, 1972). Contrast C is defined according to:

$$C = \frac{\Delta S}{S} \tag{2.17}$$

Which can be expressed as:

$$\frac{\Delta S}{S} = C > \frac{5n^{1/2}}{S} = \frac{5n^{1/2}}{n} \tag{2.18}$$

$$\therefore C > \frac{5}{n^{1/2}} \tag{2.19}$$

In order to observe a specific level of contrast C, a mean number of signal carriers n must be collected per pixel, as expressed by:

$$n > \left(\frac{5}{C}\right)^2 \tag{2.20}$$

Ultimately, since the number of electrons collected per pixel is a function of beam dwell time, contrast can be expressed in terms of the necessary beam current i_B, as follows:

$$i_B > \frac{(4 \times 10^{-18} \text{Coulombs})n_{\text{pixels}}}{\varepsilon C^2 t_f} \qquad (2.21)$$

where n_{pixels} = number of pixels in the image, ε = efficiency of signal collection per incident electron (dependent on electron generation and size/geometry of detector) and t_f = frame scan time (seconds).

Equation (2.21) is known as the *threshold equation*, defining the threshold current needed to observe a certain level of contrast for a given signal collection efficiency, enabling true sample features to be distinguished from noise in an image.

Restating Equation (2.21) with the current in terms of pA and assuming that there is one megapixel (10^6) in an image, we have:

$$i_B(pA) > \frac{4}{\varepsilon C^2 t_f} \qquad (2.22)$$

To take an example, if we have a signal generation/detection efficiency $\varepsilon = 0.25$, and we require a contrast level of at least 10 % ($C = 0.1$), with a frame time of 30 seconds, the beam current i_B would need to be a little over 53 pA. Having deduced the required beam current, this value can be used to calculate the obtainable probe size (see, for example, Goldstein *et al.*, 2003).

2.6.5 Adjusting the Contrast

The collected signal is passed to a video amplifier, the gain of which can be adjusted using the contrast setting, to stretch grey levels to fill the available range. This is called differential amplification and is shown in Figure 2.31. Note that changing the brightness of an image merely changes the DC offset (i.e. moves the trace up and down) and does not contribute to the differential signal amplification.

Figure 2.31(a) represents multiple line scans across the same part of an image, with the addition of noise, as discussed in Section 2.5.3. Figure 2.31(b) shows the effect of increasing the contrast setting: signal intensities are expanded. It is not good practice to fill the range entirely. This can result in clipping of the signal and, hence, loss of information.

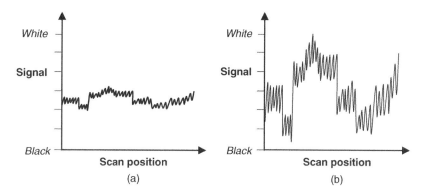

(a)   (b)

Figure 2.31 The differential video amplifier allows grey levels to be stretched, to increase the contrast and make features easier to see. This is controlled by adjusting the contrast setting on the microscope

2.6.6 Resolution

As we saw in Section 2.3, primary electrons are focused using electro-magnetic lenses and, for various reasons, the electron beam is focused into a spot of finite diameter, rather than being an infinitely sharp point. Furthermore, due to diffraction effects, each point in the image has a Gaussian profile.

A common definition for resolution is that given by the Rayleigh Criterion, which states that two points separated by a distance d_R will be resolved when their current density distributions overlap at half their separation, with an intensity drop $\leq 75\%$ of the maximum intensities (Reimer, 1985). This is illustrated in Figure 2.32.

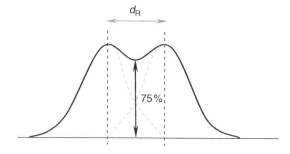

Figure 2.32 Schematic illustration of the Rayleigh Criterion for resolving two points

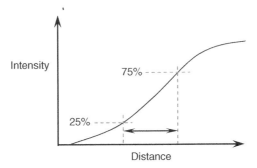

Figure 2.33 Measuring resolution based on the distance between 25 % and 75 % intensity across a sharp edge

A practical measure of resolution, often used by manufacturers as a standard for specifying the performance characteristics of a given microscope, is to determine the resolution by scanning the beam perpendicular to a sharp edge. The intensity profile will be a smoothed step function, and the resolution is usually defined as the distance measured between 25 and 75 % of the intensity profile. This is demonstrated in Figure 2.33. Note that this ratio can vary, e.g. 35:65.

In addition to the Rayleigh Criterion, there is the question of beam diameter, or spot size, relative to the size of features of interest. A large spot size works well for low magnification work but will be too large for higher magnification work if there is overlap onto adjacent pixels (see Figure 2.34), as this will cause blurring in the image. Thus, a smaller spot size is needed for high magnifications, although this brings a reduction

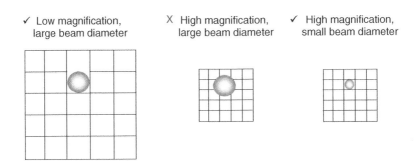

Figure 2.34 Schematic illustration to show that the beam diameter should be chosen as appropriate to the pixel size in the image. A larger beam diameter can be tolerated for low magnifications and large pixel sizes, but must be reduced when the magnification is increased and the pixel size correspondingly reduced

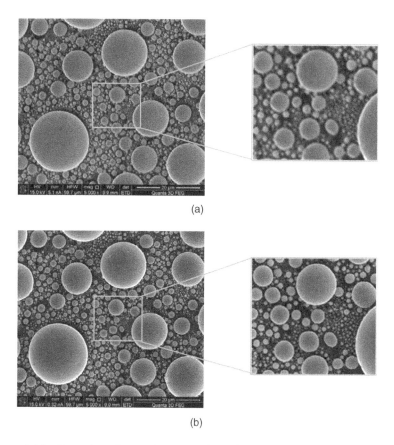

(a)

(b)

Figure 2.35 Secondary electron images to show that using a smaller spot size (beam current) gives a sharper image (b). The primary beam current i_B in (a) is 5.1 nA, while i_B in (b) is 0.32 nA. Images courtesy of Ellen Baken, FEI Company

in beam current and hence signal-to-noise ratio: images appear 'noisy', unless longer image acquisition times are used.

Meanwhile, Figure 2.35 compares images taken with two different spot sizes, showing that, in order to improve the resolution from (a) to (b), a smaller spot size (lower beam current) is needed.

2.6.7 Depth of Field

As mentioned in Section 2.1, a large depth of field is possible in SEM. This is essentially due to the fact that, above and below the focal point, the beam diameter increases gradually until features eventually become blurred. The range of the specimen that remains in focus in the

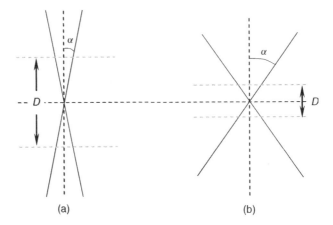

Figure 2.36 Schematic diagrams showing how depth of field D is affected by convergence semi-angle α. A smaller angle (a) gives a large depth of field, while a large angle (b) gives a smaller depth of field

z-direction defines the depth of field D. This is principally a function of the convergence semi-angle α and so, for a large depth of field, we would want to make α as small as possible, either by having a small aperture in the objective lens or by using a long working distance WD. A comparison between large and small values of α and the effect on the depth of field can be seen in Figure 2.36.

The depth of field effect is often most noticeable on tilted samples as these naturally include a range of values for working distance across the area being imaged. If the image is focused at the central plane of the specimen, this can result in blurring towards the upper and lower extremes of the image. To compensate for this, microscopes often provide a feature called *dynamic focus*. In this mode, the focus is automatically changed as the image is scanned from top to bottom, based on the angle of sample tilt. Hence, the focus is adjusted to match the change in working distance over the same range of the specimen and an improved image is obtained.

2.6.8 Image Capture

Over the past 10–15 years, the recording of images on photographic film in SEM has been replaced by digital image capture. This enables the use of different filtering techniques that can reduce noise in an image at the time of acquisition. The two most common are integration and averaging, and these are very briefly outlined below. Note that, when publishing or presenting images, it is good practice to state any image-enhancing processes that have been applied.

2.6.8.1 Integration

Live digital images can be placed in a framestore and then integrated to form a single image.[11] The process of integration smoothes out noise by adding a specified number of successive frames together (2, 4, 8, 16, 32, 64, etc). Constant signal detail is added together, while random fluctuations (noise) tend to cancel out. Integration is usually performed at fast frame rates and/or small beam currents, which is also an excellent way of avoiding artefacts arising from charge accumulation at slow frame rates and/or high beam currents. Note that integration is not useful if either the sample or the beam is drifting, since this causes image blurring.

2.6.8.2 Averaging

Also known as a recursive filter, this method increases the signal-to-noise ratio by giving a running average of several frames in real time, with pixels weighted so that the most recent frame contributes the most. Unlike integration, averaging does not stop after a specified number of frames, i.e. it is a continuous, real-time process, which means that contrast and/or brightness can be adjusted in real time.

REFERENCES

Bethe, H. (1930). Zur Theorie des Durchgangs schneller Korpuskularstrahlen durch Materie. (Theory of Passage of Swift Corpuscular Rays through Matter.). *Ann. Phys.*, 5, 325–400.

Bishop, H. (1974). Electron-solid Interactions and Energy Dissipation, in *Quantitative Scanning Electron Microscopy*. D. Holt, M. Muir, P. Grant and I. Boswarva, Academic Press.

Bongeler, R., Golla, U., Kassens, M., Reimer, L., Schindler, B., Senkel, R. and Spranck, M. (1992). Electron–Specimen Interactions in Low-Voltage Scanning Electron Microscopy. *Scanning*, 15, 1–18.

Cosslett, V. (1978). Radiation Damage in the High Resolution Electron Microscopy of Biological Materials: A Review. *J. Microsc.*, 113(pt. 2), 113–129.

Dekker, A.J. (1958). Secondary Electron Emission, in *Solid State Physics*. F. Seitz and D. Turnbull, Academic Press, Inc., pp. 251–311.

Drescher, H., Reimer, L. and Seidel, H. (1970). Backscattering And Secondary Electron Emission of 10–100 keV Electrons and Correlations to Scanning Electron Microscopy. *Z. Angewandte Physik*, 29(6), 331–338.

---

[11] It is also worth noting that the use of framestores makes digital video capture very easy.

Drouin, D., Couture, A.R., Joly, D., Tastet, X. and Aimez, V. (2007). CASINO
   V2.42 – A fast and easy-to-use modeling tool for scanning electron microscopy
   and microanalysis users. *Scanning*, 29, 92–101.
Glaeser, R. (1975). Radiation Damage and Biological Electron Microscopy, in
   Physical Aspects of Electron Microscopy and Microbeam Analysis. B. Siegel, pp.
   231–245.
Goldstein, J., Newbury, D., Echlin, P., Joy, D., Romig Jr., A., Lyman, C., Fiori, C.
   and Lifshin, E. (1992). *Scanning Electron Microscopy and X-Ray Microanalysis*,
   second edition, Plenum.
Goldstein, J., Newbury D., Joy, D., Lyman, C., Echlin, P., Lifshin, E., Sawyer, L.
   and Michael, J. (2003). *Scanning Electron Microscopy and X-Ray Microanalysis*,
   third edition, Plenum.
Goodhew, P.J., Humphreys, F.J. and Beanland, R. (2001). *Electron Microscopy and
   Analysis*, third edition, Taylor and Francis.
Hovington, P., Drouin, D. and Gauvin, R. (1997). CASINO: A New Era of Monte
   Carlo Code in C Language for the Electron Beam Interaction – Part I: description
   of the programme. *Scanning*, 19, 1–14.
Isaacson, M. (1975). Inelastic Scattering and Beam Damage of Biological Molecules,
   in *Physical Aspects of Electron Microscopy and Microbeam Analysis*. B. Siegel,
   pp. 247–258.
Jenkins, R. and Trodden, W. (1965). *Electron and Ion Emission from Solids*,
   Routledge and Kegan Paul.
Joy, D. (1995). A Database on Electron–Solid Interactions. *Scanning*, 17, 270–275.
Joy, D.C. and Joy, C.S. (1998). A Study of the Dependence of E2 Energies on Sample
   Chemistry. *Microsc. Microanal.*, 4(5), 475–480.
Joy, D.C. and Luo, S. (1989). An Empirical Stopping Power Relationship for
   Low-Energy Electrons. *Scanning*, 11(4), 176–180.
Kaneko, T. (1990). Energy Distribution of Secondary Electrons Emitted from Solid
   Surfaces Under Electron Bombardment. I. Theory. *Surf. Sci.*, 237, 327–336.
Newbury, D.E., Joy, D.C., Echlin, P., Fiori, C.E. and Goldstein, J. (1986).
   Advanced Scanning Electron Microscopy and X-Ray Microanalysis, Kluwer Aca-
   demic/Plenum Publishers.
Oatley, C.W. (1972). *The Scanning Electron Microscope, Part 1, The Instrument*,
   Cambridge University Press.
Parsons, D. (1975). Radiation Damage in Biological Materials, in *Physical Aspects
   of Electron Microscopy and Microbeam Analysis*, B. Siegel, pp. 259–265.
Reimer, L. (1975). Review of the Radiation Damage Problem of Organic Speci-
   mens in Electron Microscopy, in *Physical Aspects of Electron Microscopy and
   Microbeam Analysis*. B. Siegel, pp. 231–245.
Reimer, L. (1985). *Scanning Electron Microscopy. Physics of Image Formation and
   Microanalysis*, Springer-Verlag.
Reimer, L. and Schmidt, A. (1984). The Shrinkage of Bulk Polymers by Radiation
   Damage in an SEM. *Scanning*, 7, 47–53.
Seiler, H. (1983). Secondary Electron Emission in the Scanning Electron Microscope.
   J. Appl. Phys., 54(11).
Talmon, Y. (1987). Electron Beam Radiation Damage to Organic and Biologi-
   cal Cryo-specimens, in *Cryotechniques in Biological Electron Microscopy*, R.A.
   Steinbrecht and K. Zerold., Springer-Verlag.

3

General Principles of VP-ESEM: Utilising a Gas

3.1 INTRODUCTION

In VP-ESEM, interactions of electrons in the gas and the subsequent formation of positive ion by-products are particularly useful in allowing insulating specimens to be imaged without the need for a conductive coating. However, there are many interdependent parameters involved and we must always consider how a change in even one parameter will affect resultant images or spectra.

Broadly speaking, the aims of this chapter are to look at the ways in which gases are used in the VP-ESEM, starting with an introduction to the instrument and general principles in Section 3.2. Excitation and ionisation of gas atoms or molecules and the movement of both positive and negative charge carriers are processes of great importance in providing detectable signals in VP-ESEM. Hence, the potential for generating image-forming signals using electrons, photons and ions will be explained in Section 3.3.

Meanwhile, air and water vapour are commonly used as imaging gases in VP-ESEM, and water vapour is particularly useful as it can be used to control the thermodynamic stability of moist or liquid specimens as well as having a role in dynamic hydration and dehydration experiments. Hence, the properties and potential of the use of water vapour will be covered in detail in Section 3.4. Other gases may be selected for a variety of reasons, usually because the specimen is sensitive to certain gases or because the conditions of a given experiment dictate the use of a specific

Principles and Practice of Variable Pressure/Environmental Scanning Electron Microscopy (VP-ESEM)
D. J. Stokes
© 2008 John Wiley & Sons, Ltd

gas. We will deal with some of these alternative gases and their effects in Chapter 4 and look at some of their applications in Chapter 6.

3.2 VP-ESEM INSTRUMENTATION

3.2.1 Typical Features

A VP-ESEM instrument typically has the components shown in Figure 3.1. Many of the features are just the same as for high-vacuum SEM and, indeed, the modern VP-ESEM has similar capabilities when used in high-vacuum mode. However, the ability to maintain a gaseous chamber environment does call for a few technological differences when operating in VP-ESEM mode. For the background history on the development of this technology, refer to Chapter 1.

One distinguishing feature of the VP-ESEM over the conventional high-vacuum SEM is the presence of differentially pumped zones, separated by pressure-limiting (or differential) apertures. Note that the

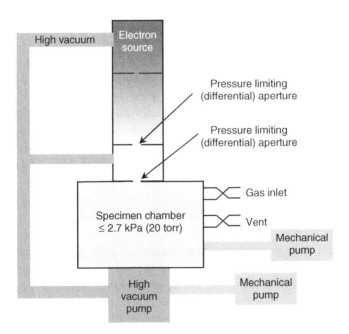

Figure 3.1 Simplified schematic diagram to show the arrangement of zones and pressure-limiting (differential) apertures of a typical VP-ESEM instrument (not to scale)

electron source is maintained under high vacuum, according to the type of filament, just the same as for conventional SEM (Chapter 2, Section 2.2). Similarly, the main part of the electron column is under high vacuum, to minimise primary electron scattering as far as possible. However, the pressure in the specimen chamber can be anything up to 2660 Pa (20 torr): a significant departure from the conditions of a high-vacuum SEM. Hence, throughout Chapters 3 and 4 we consider the implications this has for imaging and analysis.

In addition to differential pumping, operating in VP-ESEM mode often necessitates the use of detectors capable of operating in a gas, since the traditional Everhart–Thornley secondary electron detector (ETD) would cause arcing in a gaseous environment.[1] The various methods of collecting or detecting signals in the gas generally rely on some form of pre-detection amplification of the signal, in contrast to the ETD in which collected electrons give rise to a shower of photo-electrons, post acquisition (see, for example, Reimer, 1985 and Goldstein *et al.*, 2003).

3.2.2 Primary Electron Scattering in VP-ESEM – the General Case

The suitability of different gases for imaging and analysis will be discussed quantitatively in Chapter 4. First we qualitatively consider the general effect that a gaseous environment has on the transit of primary electrons *en route* to the specimen.

The progress of a primary electron is a function of its mean free path λ or, to put it another way, the average distance travelled before colliding with a gas molecule. Figure 3.2(a) shows schematically that in a high-vacuum environment, there is the occasional chance of a random scattering event to knock a primary electron from its intended trajectory. In the case where there are some gas molecules in the specimen chamber, shown in (b), there is a degree of scattering (termed oligo-scattering), but a focused central beam still makes it to the specimen surface. At higher pressure, primary electrons become scattered completely (plural scattering), as shown in Figure 3.2(c).

Correspondingly, Figure 3.3 shows the resultant primary beam profiles in each case. This is a way of representing the instantaneous intensity of primary electrons at each point along the path across the specimen. The focused primary beam in high vacuum would have the profile of a

---

[1] The Everhart–Thornley detector uses a bias in the order of several thousand volts on the scintillator.

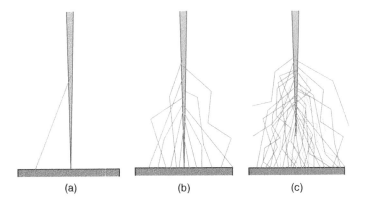

Figure 3.2 Generalised schematic of the scattering of an electron beam in vacuum and through a gas. In (a), the high-vacuum case, the primary beam is largely unscattered. In (b), where there are gas molecules in the chamber, some scattering occurs (oligo-scattering). Crucially, there is a sufficiently unscattered part of the primary beam to form a focused probe. If the pressure is much too high, as in (c), primary electrons are completely scattered (plural scattering) and do not form a focused probe

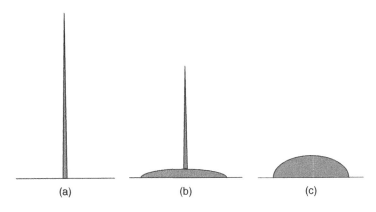

Figure 3.3 Idealised primary beam intensity profiles as a result of (a) minimal scattering, (b) oligo-scattering and (c) plural scattering

focused beam as shown in (a) (the Gaussian distribution has been omitted for simplicity), while the oligo-scattered beam (b) has superimposed on it what is known as a 'skirt' of scattered electrons. With plural scattering (c), it is not possible to form a coherent beam profile. Hence, electron microscopy is not performed at atmospheric pressure, for example.

The region where primary electrons are most at risk of being scattered to form a skirt is once they exit the final pressure-limiting (differential) aperture close to the specimen surface. Note that, in general, the central, focused part of the beam in VP-ESEM, whose profile is depicted in Figure 3.3(b), is still just as good as the focused beam in the 'no scattering' case shown in Figure 3.3(a). The main difference is that some primary electron beam current is lost into the skirt, yielding a lower signal from the impact point of the beam.

Crucially, the resolution – the size of features that can be visualised, as defined by the beam diameter – is not ordinarily reduced relative to the high-vacuum case. The skirt electrons do, of course, result in emission of signals from all around the area of interest and thus add to the background signal. Figure 3.4 is a comparison between imaging with VP-ESEM in high vacuum and in the presence of a gas.

As we will see in Chapter 4, the size of the beam skirt can be very significant, depending on the chamber gas and a number of other parameters. Indeed, the beam skirt can fill the entire field of view, particularly at higher magnifications, regardless of where the primary beam is at any given instant. In general, this means that the skirt-induced signal does not particularly vary with primary beam position. This background signal adds a DC offset and increases the statistical noise in the information-carrying signal (see Chapter 2, Section 2.6.5). The latter reduces the signal-to-noise ratio S/N, making it more difficult to distinguish features having small contrast differences (Chapter 2,

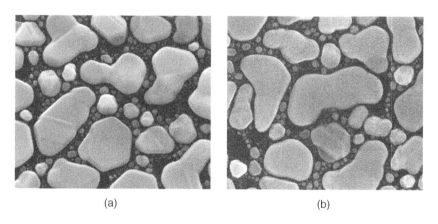

(a)   (b)

Figure 3.4 Images of gold on carbon for comparison between imaging in high vacuum (a) and in a gas (b). The image in (b) was obtained using 600 Pa (4.5 torr) of water vapour. In (b) there is no loss of resolution, but a reduction in signal-to-noise ratio. Primary beam energy $E_0 = 20$ keV. Horizontal field width $= 1.3$ μm. Images courtesy of Daniel Phifer, FEI Company

Table 3.1 Common units of pressure and their conversions into different units

Atmospheric pressure = 760 torr (or mmHg) = 101×10^3 Pa = 1 bar		
1 torr	~ 133 Pa	~ 1.33 mbar
1 Pa	$\sim 7.5 \times 10^{-3}$ torr	~ 0.01 mbar
1 mbar	~ 0.75 torr	~ 100 Pa

Sections 2.6.3 and 2.6.4). This is compounded by the fact that the *S/N* is reduced due to loss of current in the focused beam.

However, despite these apparent difficulties, high-resolution, high-quality images can be obtained in the VP-ESEM under oligo-scattering conditions, especially if filtering techniques such as integration or averaging are used to improve image quality (Chapter 2, Section 2.6.8). Alternatively the signal can be boosted by increasing the primary beam current (which tends to enlarge the beam diameter, sacrificing resolution at high magnification) or by increasing the dwell time of the electron beam (but with the increased risk of localised charging and radiation damage for sensitive specimens).

3.2.3 Units of Pressure

The SI unit of pressure is the pascal (Pa), but units such as torr or millibar (mbar) are in common use in the literature. The conversions given in Table 3.1 will therefore be helpful.

Graphical data in this chapter are plotted in units of both Pa and torr, for ease of reference. To convert to mbar, simply divide the values in Pa by a factor of 100.

3.3 SIGNAL GENERATION IN A GAS

3.3.1 Introduction

In Chapter 1 it was noted that there are several methods for the generation and collection of signals in VP-ESEM. In the following sections, we see how those ideas are put into practice. The use of gases in the specimen chamber is the main differentiator between SEM and VP-ESEM, and typical gases include air, water vapour, helium, argon, nitrous oxide, carbon dioxide and nitrogen. Here we will look at the general physical properties of gases, their interactions with electrons and some of the electromagnetic species that can be generated as a result. In Chapter 4 we will deal with the properties of some specific gases as a function of imaging parameters in VP-ESEM, to gain a more quantitative understanding.

3.3.2 Direct Collection of Electrons and Ions

3.3.2.1 Ionised Gas Cascade Signal Amplification

The simplified arrangement shown in Figure 3.5 helps to demonstrate how an electric field[2] can be set up between a biased electrode and a conductive specimen placed on a grounded specimen stage.

Note that the working distance WD is defined as the distance from the objective lens to the specimen, and is shown on the microscope display, whereas the anode–specimen d distance will differ from WD by an amount equal to the distance from the end of the objective lens to

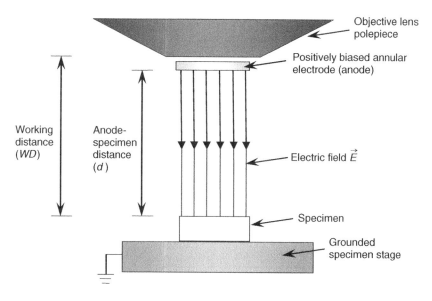

Figure 3.5 Simplified schematic diagram to illustrate the electric field that develops between two electrodes, one of which is negatively biased and the other at ground potential (0 volts). The spacing between the lines tells us the relative electric field strengths as a function of position (a uniform field, in this case)

---

[2] By convention, the arrows denoting an electric field are taken to point in the direction in which positive charge would move.

the pressure-limiting (differential) aperture, the latter often being within the anode assembly (where applicable).

Using such an arrangement in the VP-ESEM, a relatively small positive potential applied to the anode causes secondary electrons emitted by the specimen to be accelerated away from the specimen surface. These secondary electrons collide with and ionise gas molecules in their path. Further electrons are produced as a result, and these too can participate in ionising collisions, thus propagating the cascade towards the anode and amplifying the secondary electron signal.

We can therefore take advantage of the presence of a gas in the sample chamber by collecting the low-energy electrons that have been liberated in the gap between the specimen and anode. The underlying principles are akin to the Townsend gas capacitor model (see, for example, von Engel, 1965; Moncrieff *et al.*, 1978; Danilatos, 1990b; Durkin and Shah, 1993; Meredith *et al.*, 1996; Thiel *et al.*, 1997 and Thiel, 2004). The amplification process is shown schematically in Figure 3.6.

Unfortunately, it is not possible to legislate for the cascade and collection of various unwanted electron signals, such as those generated by collisions of primary or backscattered electrons with gas molecules,

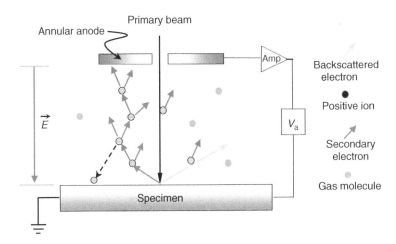

Figure 3.6 A simplified schematic diagram of gaseous signal amplification via ionisation. The primary electron beam impinges on the sample, leading to the production of backscattered and secondary electrons. The electric field between the positively biased anode and the grounded specimen stage accelerates secondary electrons towards the anode. Secondary electrons collide with and ionise gas molecules in their path and amplify the secondary electron signal. Positive ions drift towards the sample surface to alleviate negative charge build up

or different types of secondary electron signal, such as those generated by backscattered electrons exiting the specimen surface (SE$_{II}$) or emitted as a result of wayward primary electrons striking the chamber walls, polepiece, etc. (SE$_{III}$). Thus, the background signal is also amplified. Work by Moncrieff *et al.* (1978), Meredith *et al.* (1996), Fletcher *et al.* (1997) and Thiel *et al.* (1997) gives a detailed treatment of the factors involved in signal amplification in VP-ESEM by the cascade mechanism. The emphasis is on developing a systematic approach for optimising image quality for a given set of experimental conditions, such as gas pressure, gas type and the anode–specimen distance d.

The amount by which the signal of each emitted electron is multiplied in the cascade process is known as the gain. Assuming that the gain in electron signal is analogous to an equal but opposite positive ion current I_+, the gain can be experimentally determined by measuring the positive ion current for a given set of conditions (see Meredith *et al.*, 1996; Toth and Phillips, 2000 and Thiel, 2004). The idea is basically this: the primary electron beam current I_0 defines the total current in the system, and this can be measured by using a Faraday cup. Various signal components arise in the gas in response to I_0, yielding an amplified electron current I_{amp} as well as the corresponding positive ion current. The amplified gaseous electron current is a composite of the contributions to ionisation from primary, backscattered and secondary electrons, whose currents can be denoted $I_{g(PE)}$, $I_{g(BSE)}$ and $I_{g(SE)}$, respectively:

$$I_{amp} = I_+ = I_{g(PE)} + I_{g(BSE)} + I_{g(SE)} \qquad (3.1)$$

The amplification coefficient for a given gas is then found simply from the ratio of positive ion current to primary beam current, I_+/I_0. Recall that the specimen itself determines the magnitude of the electron emission coefficients η and δ (see Chapter 2), and so gaseous signal amplification is inherently specimen-dependent.

Assuming a uniform electric field[3] and steady-state conditions, the amplification of electron signals in the VP-ESEM can be described by the following expression derived by Thiel *et al.* (1997):

$$I_+ = I_0 k (e^{\alpha_\infty d} - 1) \left\{ S_{pe} \frac{p}{\alpha_\infty} + \delta + \eta S_{bse} \frac{p}{\alpha_\infty} + \delta_{se2} \eta \right\} \qquad (3.2)$$

---

[3] This turns out to be an oversimplification, since VP-ESEM detector fields are quite often not uniform, as later shown by Toth *et al*, for example.

where α_∞ = steady-state ionisation efficiency of the gas (ion pairs mm^{-1}), d = sample–anode gap, I_0 = primary beam current, S_{pe} = ionisation efficiency of primary electrons (ion pairs $\cdot$ mm^{-1} $\cdot$ torr^{-1}), S_{bse} = ionisation efficiency of backscattered electrons (ion pairs $\cdot$ mm^{-1} $\cdot$ torr^{-1}), η = backscattered electron coefficient, δ = secondary electron coefficient, p = chamber gas pressure and $k = a$ gas-specific amplification factor related to inelastic scattering cross-sections.

The factor α is called Townsend's first ionisation coefficient, which is given by:

$$\alpha = Ape^{-Bpd/V_a} \tag{3.3}$$

where, again, A and B are gas-specific constants, p is pressure, d is the specimen–anode distance and V_a is the anode bias. Hence, signal amplification is also dependent on the bias applied to the anode.

To simplify the discussion, we can think of the ionisation efficiency α for a given gas as being proportional to the pressure–distance product pd and inversely proportional to the anode bias V_a.

In addition, the signal gain g is a function of the ionisation coefficient α and the specimen–anode distance d, such that:

$$\ln(g) = \alpha d = Apde^{-Bpd/V_a} \tag{3.4}$$

Which can be written:

$$g = e^{\alpha d} \tag{3.5}$$

Equations (3.3)–(3.5) have been used to show how the relative gain in signal varies as a function of anode bias V_a, for a fixed value of the pressure–distance product pd (Thiel, 2004). This is significant: it tells us that if we wish to maintain a constant signal over a range of different pressures, for example, then the specimen–anode distance must be adjusted accordingly to keep the value of pd constant. In other words, if the pressure p is increased, the distance d should be decreased to compensate, and vice versa. This does neglect one or two complicating factors, but nonetheless provides a useful basis on which to begin understanding differences in signal gain characteristics for a given specimen–gas system as a function of just these few parameters.

Figure 3.7 is an example plot, obtained using Equation (3.2), to show the various contributions to the total signal made by backscattered electrons (BSEs), primary electrons (PEs) and secondary electrons (SEs)

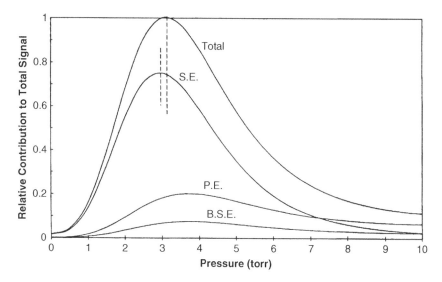

Figure 3.7 Graph of signal contributions as a function of chamber gas pressure (water vapour) for a carbon substrate for a specific set of working conditions (see text for details). The maximum total amplification occurs at slightly higher pressure than the secondary electron maximum, as shown by the dashed lines. At high pressures, the signal is dominated by background signals such as ionisation of gas molecules by primary electrons. Reproduced from Thiel *et al.* (1997), copyright Wiley-Blackwell

for a carbon substrate (anode bias $V_a = 300$ V, specimen–anode distance $d = 7$ mm) (Thiel *et al.*, 1997).

It can be seen that the position of the peak for total amplification is different to that for the amplification of secondary electrons. In principle, by operating the microscope just below the maximum in the total amplification peak, a purer secondary electron signal can be obtained. It can also be seen that background signals dominate at higher pressures.

3.3.2.2 The Specimen–Anode Gap

Secondary electrons are typically emitted with just a few electronvolts of energy, and so need to be accelerated in the electric field between the specimen and anode before acquiring the energy needed to cause ionisation of gas molecules. Assuming a constant electric field, this creates an avalanche of amplified secondary electrons, the threshold energy for which defines what are known as steady-state 'swarm' conditions that must be achieved in initiating the gas cascade (Boeuf and Marode,

1982; Thiel *et al.*, 1997). This is related to the ionisation potential of the specific gas being used.

However, it takes time to accelerate the emitted secondary electrons, so if the specimen–anode gap is too small, secondary electrons arrive at the anode too soon – before the threshold conditions for amplification have been met – and so the amplification process will not get started. This is why it is essential to have at least some space between the specimen and an electron–gas cascade secondary electron detector. The minimum gap distance should be at least the diameter of the final pressure-limiting (differential) aperture, to avoid the effects of low pressure due to aperture pumping in this region.

3.3.2.3 *The General Shape of the Amplification Curve*

The amplification of electron signals has a characteristic shape, as depicted in Figure 3.7, for the reasons outlined below.

The detection of the secondary electron signal is very sensitive to the pressure of the gas. Since secondary electrons have inherently low energy, the mean free path λ of a typical secondary electron is quite short (on the order of a few tens to a few hundred microns). Note that, in Chapter 4, we will make some quantitative analyses of the pressure dependence of primary electron mean free paths. For now, we are interested in the behaviour of secondary electrons in the context of signal detection.

If the pressure is high, and hence the concentration of gas molecules is large, then the mean free path is very short and a significant fraction of secondary electrons is scattered before they acquire sufficient energy to cause ionisation. As the pressure is reduced, the mean free path increases, the average kinetic energy of the secondary electrons increases and so signal amplification becomes more efficient. Eventually, though, if the concentration of gas molecules is too small, amplification events are few and far between, and so the secondary electron signal decreases. Concurrently, this can result in there being insufficient ions to control the charge state of the specimen.

3.3.3 Collection of Photons – the Gas Luminescence Signal

3.3.3.1 *Photon Production*

As we know from the previous section, when an electron collides with an atom or molecule some of its energy can be transferred and this can result in ionisation. However, this is not the only inelastic process that interests us in VP-ESEM. Extra energy can be accommodated in a number of ways, one of which is for an electron in an inner shell to move to a higher orbital in a process called excitation. After a short time, the atom or molecule relaxes back to its neutral ground state: the electron drops back down to its original shell. As there is a specific energy difference for the transition between orbitals, excess energy is expended by the emission of a photon of light. This is analogous to the process of cathodoluminescence shown in Chapter 2, Section 2.4.5. Figure 3.8 shows the relationship between excitation and ionisation.

In combination with primary, backscattered and secondary electrons (see Figure 3.9), there is scope for many of these excitation events in the presence of a gas. Indeed, an appreciable amount of light can be produced in the ultraviolet, visible and infrared parts of the electromagnetic spectrum. This is known as gas luminescence, or scintillation, and is another mechanism by which a signal is formed in the VP-ESEM, as described in this context by Danilatos (1986). Excited states typically have short lifetimes, on the order of nanoseconds, so that relaxation quickly follows and photons are produced rapidly. Morgan and Phillips

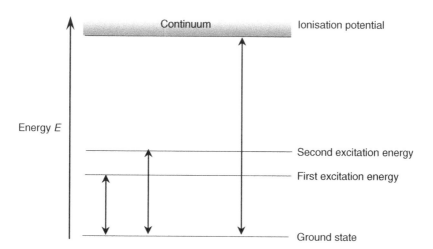

Figure 3.8 A highly simplified schematic diagram showing possible transitions between the ground state and an excited state, compared to ionisation which involves removal of the electron from the atom. An electron can be excited to a higher energy level, and when the atom returns to the ground state, a photon is emitted

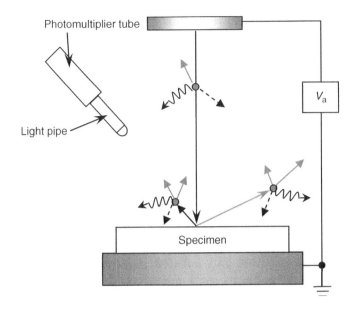

Figure 3.9 Collecting the gas luminescence signal generated by excitation–relaxation and electron–ion recombination of atoms or molecules in a gaseous environment. The events are caused by primary, backscattered and secondary electron interactions, each of which is depicted in the diagram, using the same key as in Figure 3.6. Adapted from Morgan and Phillips (2006)

(2006) provide an excellent overview of the features of this type of signal generation and collection, including a quantitative analysis of the photon yield as a function of pressure, working distance and anode bias.

Another way in which photons can be produced is through electron–ion recombination. This is not generally of the sort where an ion recaptures its newly liberated electron, since electrons and ions are quickly swept apart in the electric field between the anode and the specimen. More likely, electron–ion recombination occurs as ions capture electrons emitted from the specimen or produced in the gas cascade. More will be said about this in Chapter 5.

Photons that have been generated via excitation–relaxation or electron–ion recombination can be collected using a light pipe and guided into a photomultiplier. This converts photons into pulses of electrons, which can then be electronically amplified, thereby increasing the intensity of the signal. Figure 3.9 schematically shows a typical arrangement for collecting these photons in the VP-ESEM.

The overall photon 'cascade' signal[4] A_{hv} per incident electron can be described by Equation (3.6) (Morgan and Phillips, 2006):

$$A_{hv} = \frac{\alpha_{exc}[e^{\alpha_{ion}d} - 1](S_{PE}p + \eta S_{BSE}p + \delta)}{\eta + \delta} + \frac{pd(\omega_{PE} + \omega_{BSE})}{\eta + \delta} \quad (3.6)$$

where α_{exc} = excitation coefficient, α_{ion} = Townsend's first ionisation coefficient (analogous to α_∞ in Equation (3.2)), d = distance, S_{PE} and S_{BSE} = ionisation efficiencies of primary and backscattered electrons, respectively, ω_{PE} and ω_{BSE} represent photon amplification efficiencies of primary and backscattered electrons, respectively, and η and δ are the secondary and backscattered electron coefficients, respectively

The first term in Equation (3.6) is the scintillation caused by cascading secondary electrons, while the second term is amplification due to primary and backscattered electrons. From this, we can see that using the gas luminescence signal for imaging does not necessarily entail ionising collisions in the gas nor, hence, the generation of secondary electron-induced positive ions. By keeping the acceleration of secondary electrons just low enough, their energies can be tailored to remain below the ionisation threshold, so that the first term in Equation (3.6) is effectively zero. In that case, signal generation is independent of the cascade mechanism and is essentially a function of electron excitation and relaxation events generated by both primary and backscattered electrons in the gas: the second term in Equation (3.6).

In the absence of any significant secondary electron cascade and ion production in the gas, the charge-suppression mechanism can proceed on the basis of just a small but adequate number of ions generated by collisions of primary and backscattered electrons with gas molecules. This is an important point. It turns out that a high concentration of ions can be disadvantageous, as will be discussed in Chapter 5, and so there could be benefits to using the gas luminescence mechanism to avoid this situation while generating an imaging signal.

3.3.3.2 Enhanced Photon Signals

The photon signal is normally omnidirectional, i.e. not favouring any particular direction, which means that some signal is lost because the photons do not happen to be in the vicinity of the detector. In addition,

---

[4] This is not to say that the photons themselves are amplified, but rather that they are a by-product of the electron cascade process.

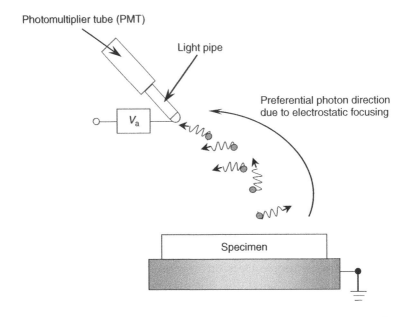

Figure 3.10   Schematic diagram showing the principle of enhanced gas scintillation. A small bias near the photomultiplier guides secondary electrons towards it and thereby helps to give a directional and more intense gas luminescence signal in the gas along the path followed by the charged particles

the efficiency of the luminescence process decreases with decreasing electric field strength.

However, locating a positive bias near to the photomultiplier collector grid has been shown to have the beneficial effect of creating a more intense, localised electric field in the gap between the specimen and the detector (Morgan and Phillips, 2006). This increases the luminescence efficiency near the light pipe, as well as attracting the secondary electron cascade along a specific path, thereby increasing photon signal collection in that direction. Figure 3.10 illustrates this point.

3.3.4 Detecting Indirect Electron and Ion Currents

3.3.4.1 Charged Signal Carriers and Induced Currents

The principle of induction can be invoked to describe an alternative method of detecting signals in VP-ESEM. Simply put, when a charge carrier moves between two conductors, an electric charge is induced

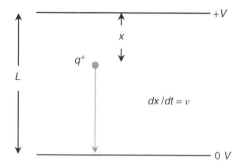

Figure 3.11   Schematic diagram showing a charge *q* (positive in this case) moving between two electrodes a distance *L* apart. The distance moved with time, d*x*/d*t* determines the velocity *v* of the moving charge between the electrodes

on the conductors. If these conductors are connected together, then a current will flow between them.

We can extend this idea if we now consider the familiar arrangement of a positively biased electrode (anode) placed near the specimen surface, with the specimen in contact with the stage, which acts as a second electrode. Clearly, negative charge carriers (e.g. emitted secondary electrons) will be attracted towards the anode and positive charge carriers (e.g. ions generated in the gas cascade process) will be repelled. Figure 3.11 shows a positively charged particle moving between two electrodes, where the velocity, *v*, of the particle is determined by the distance *d* travelled per unit time *t*, i.e.:

$$dx/dt = v \qquad (3.7)$$

Note that the drift velocity of ions is dependent on gas type, electric field strength and pressure, but is typically on the order of $10^2 \, ms^{-1}$. The effects of this (slow) drift rate will come up again in Chapter 5.

According to the induction principle, then, the net movement of these charge carriers in the gas induces a pulse of current that is felt at the electrodes, and by placing a specimen current amplifier at the specimen, the current induced by the movement of the ions can be collected. When the charge carriers cease movement (due to recombination or termination at an electrode), their contribution to the induced current also ceases. So, there is no requirement for the charge carriers to actually arrive at either electrode: their mere approach induces a current and, by collecting this current, the moving charge carriers become charged signal carriers. This is the principle of the biased electrode method described by

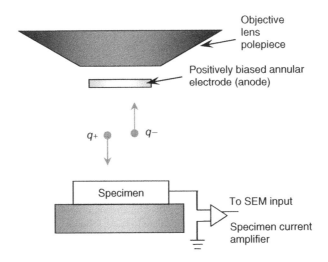

Figure 3.12   Collecting an induced current at the specimen stage due to the presence of moving charge carriers in the gas

Danilatos (1990a), Farley *et al.* (1990), Farley and Shah (1991), Mohan *et al.* (1998) and Morgan and Phillips (2005), as shown in Figure 3.12. Induced currents have also been implicated as components to the total signal for anodes collecting the direct secondary electron signal (Phillips *et al.*, 1999).

The magnitude of the induced current $i_{induced}$ in the VP-ESEM can be estimated from Equation (3.8) (see Mohan *et al.*, 1998):

$$i_{induced} = \frac{q}{L}.v \tag{3.8}$$

where q is the number of charge carriers, L is the distance between the electrodes and v is the velocity of the charged particles.

This arrangement works for both conducting and nonconducting specimens. We will defer discussion of the material properties of non-conductive materials until Chapter 5. For the moment, it is noted that, for a nonconductive (dielectric) material, an electric field can be supported within the specimen, allowing a current to be induced. Hence, the charge carriers do not have to physically pass through a nonconductive material (which of course would be very difficult in an electrical insulator, by definition). Positive ions serve a dual purpose in this case, both acting as signal carriers and helping to mask negative specimen charging at the surface.

As with the other methods of detection discussed, changing the gas pressure or the anode potential has an effect on the induced current signal, as we might expect (see, for example, Mohan *et al.*, 1998).

3.4 IMAGING WITH WATER VAPOUR

3.4.1 Introduction

When the imaging gas is water vapour, there are numerous experiments that can be performed that are unique to VP-ESEM. For example, it is possible to maintain an environment corresponding to the saturated vapour pressure of water (100 % relative humidity, RH) around a moist specimen, or an equilibrium vapour pressure appropriate to that specimen, so that the specimen is neither hydrating nor dehydrating. Saturated vapour pressure varies as a function of temperature, and it is quite convenient that between 0 °C and 25 °C, values for saturated water vapour pressure occur at pressures between 600 Pa and 2.66 kPa (4.5 to 20 torr).

These criteria are suitable for stabilising and imaging liquid-containing specimens, particularly those of a biological nature, in the VP-ESEM. Alternatively, experiments can be carried out in which water is added or removed from the specimen, perhaps to enable observation of a chemical reaction or change of state.

Some specific applications involving imaging with water vapour will be given in Chapter 6. This section explains the physical principles needed in order to ensure that the correct conditions are chosen for a given experiment.

3.4.2 Thermodynamic Equilibria

3.4.2.1 Pure Water

Water molecules can escape across a liquid–air interface, transported by diffusion and convection, and some proportion of these molecules will inevitably return to the liquid. In a closed system, the exchange of water molecules between liquid and vapour eventually settles down to a thermodynamic equilibrium between the two phases. Hence, evaporation and condensation occur at equal rates. Under equilibrium conditions at a given temperature, there is a specific amount of vapour above the liquid, described as the saturation vapour concentration or, equivalently, saturated vapour pressure.

Thermodynamic theory, embodied in the Clausius–Clapeyron equation (Equation (3.9)), can be used to plot the phase behaviour of a liquid as a function of temperature and pressure:

$$\frac{dp}{dT} = \frac{L}{T} \frac{1}{V - V_L} \tag{3.9}$$

where p = pressure, T = temperature, L = latent heat of vaporisation per mole, V = volume occupied by one mole of the vapour and V_L = volume occupied by one mole of the liquid (see, for example, Tabor, 1991).

Figure 3.13 shows part of the phase diagram for water. Points that lie on the curve represent thermodynamic equilibria – water molecules are evaporating and condensing all the time, but the net liquid–vapour ratio remains constant for a given temperature. This is the saturated vapour pressure, SVP, of the liquid. Equally, such a plot can be produced from tables of experimental values, some of which are shown for water in Table 3.2 for convenience (Lide, 1991).

Temperatures and pressures can be adjusted in order to attain equilibrium or nonequilibrium conditions, as required. A nonequilibrium state means that the concentration of vapour molecules above the specimen

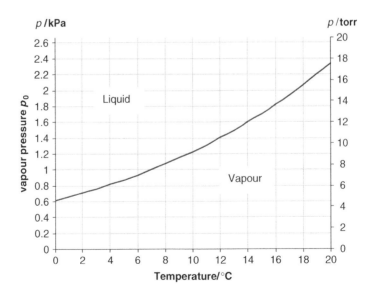

Figure 3.13 Plot of saturated vapour pressure of pure water. The curve represents the temperatures and pressures required so that 100 % relative humidity is achieved, such that the net loss/gain of water is zero

Table 3.2 Values of temperature and pressure corresponding to the saturated vapour pressure of pure water (Lide, 1991)

Temperature/°C	Temperature/K	Pressure/kPa	Pressure/torr
0	273	0.611	4.58
1	274	0.657	4.93
2	275	0.706	5.30
3	276	0.758	5.60
4	277	0.814	6.10
5	278	0.873	6.55
6	279	0.935	7.01
7	280	1.002	7.52
8	281	1.073	8.05
9	282	1.148	8.61
10	283	1.228	9.21
11	284	1.313	9.85
12	285	1.403	10.52
13	286	1.498	11.23
14	287	1.599	11.99
15	288	1.706	12.80
16	289	1.819	13.64
17	290	1.938	14.54
18	291	2.064	15.48
19	292	2.198	16.49
20	293	2.339	17.54

is either higher or lower than that required for a stable state. This will lead to an imbalance in the exchange of molecules between the liquid and the vapour. A higher concentration will lead to an increase in the number of vapour molecules landing on the specimen surface (condensation), while a lower concentration will shift the balance in favour of molecules escaping the surface (evaporation). This is illustrated in Figure 3.14. Appropriate control over these properties is very useful as it enables dynamic experiments to be carried out *in situ* in the VP-ESEM.

3.4.2.2 Aqueous Phases

Many systems containing one or more aqueous phases (e.g. hydrated specimens) consist not of pure water, but of aqueous phases containing dissolved solutes. We should, therefore, consider what influence these solutes have on the vapour pressure of aqueous systems.

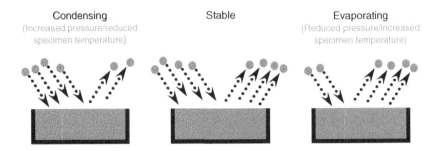

Condensing   Stable   Evaporating

(Increased pressure/reduced specimen temperature)   *(Reduced pressure/increased specimen temperature)*

Figure 3.14 Simplified schematic diagram to illustrate thermodynamic equilibrium and nonequilibrium conditions. Unless the vapour above a liquid is in balance with its liquid, condensation or evaporation will occur. Under balanced conditions (i.e. equilibrium), the exchange of molecules between the vapour and the liquid is the same and so the net difference in the number of molecules in each phase is zero

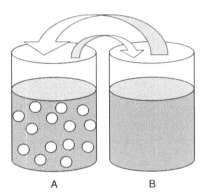

A   B

Figure 3.15 Solutes in an aqueous solution (vessel A) and pure water (vessel B). Water molecules continually exchange between the vapour and the liquid in each case. However, solutes lower the vapour pressure in A, so the liquid–vapour exchange is reduced if the vessel is in isolation. If placed together in a sealed container, water molecules from B will be driven into vessel A

According to Raoult's law, the vapour pressure of a solution is proportional to the mole fraction of solute (Tabor, 1991). An important consequence of this statement is that the vapour pressure of a solution is *less* than that of the pure solvent. The diagram in Figure 3.15 shows two vessels of water, one of which contains solute molecules, depicted by white circles. In isolation, the processes of evaporation and condensation occur in each vessel such that the vapour of each is in equilibrium with its liquid.

However, the vapour pressure and hence concentration of vapour molecules above pure water (vessel B) is higher than for the solution (vessel A). Now, if these containers are placed together in a sealed box, there will be a net evaporation from the pure water vessel and condensation into the solution vessel in an osmosis-like flow.

This driving force for solvent to enter a solution, known as osmotic pressure, was thermodynamically described by van't Hoff, and essentially depends upon the number of solute molecules contained in the solution (again, for greater detail of these concepts, refer to texts such as Tabor, 1991).

Physiological solutions tend to have large osmotic pressures, the magnitudes of which are not always adequately predicted by theory alone. This is because it is assumed that solutions are dilute (solutes occupy negligible volume) and that their behaviour is ideal (solute molecules do not interact with each other or with solvent molecules). Real physiological aqueous phases, such as those found in the interiors of mammalian cells, are neither dilute nor ideal: macromolecules such as proteins and polysaccharides take up a large volume – around 30 % of the available space – and interact strongly with water molecules (protein folding, for example, is dependent on such interactions) (Ellis, 2001).

A useful concept that takes account of the factors outlined above is that of water activity a_w, which describes the energy state of the system and is defined as the ratio of the equilibrium vapour pressure p_{eq} of the liquid or substance to the vapour pressure p_0 of pure water at the same temperature:

$$a_w = p_{eq}/p_0 \qquad (3.10)$$

If we relate the water activity of a solution to its relative humidity RH (since $RH = a_w \cdot 100$), then the saturated vapour pressure curve for water can be modified to reflect the equilibrium vapour pressure of a given aqueous phase using Equation (3.11):

$$p_{eq} = a_w p_0 \qquad (3.11)$$

where p_0 represents the vapour pressure for relative humidity RH = 100 % at a specific temperature for pure water, and a_w is the water activity of the aqueous, solute-containing phase.

If we take the water activity a_w of a saturated solution of common salt NaCl as a guide, with $a_w = 0.75$ (relative humidity RH = 75 %), it becomes clear that the extent to which this factor lowers the vapour pressure of a solution becomes quite significant. For example, the equilibrium vapour pressure for a physiological system, perhaps, at a

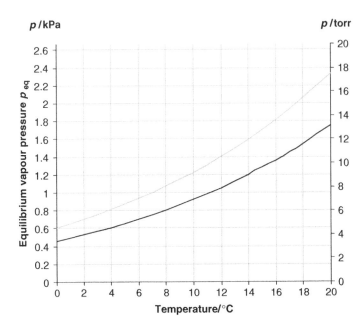

Figure 3.16 Plot of equilibrium vapour pressures as a function of temperature for a specimen having a vapour pressure equivalent to 75 % relative humidity RH (black line). Data for the saturated vapour pressure of water (i.e. 100 % RH) is also shown (light grey line)

temperature $T = 3\,°C$ would be $p_{eq} = 572\,Pa$ (4.3 torr), 25 % lower than the pressure for pure water where $p_0 = 758\,Pa$ (5.6 torr).

Using the value of a_w for salt stated above, Equation (3.11) is plotted in Figure 3.16, alongside the original data for pure water vapour (where $a_w = 1$) shown earlier in Figure 3.13.

3.4.3 Nonequilibrium Conditions

Another consideration of the behaviour of aqueous systems is what happens under nonequilibrium conditions. In particular, what are the kinetic implications: what is the rate of water loss?

The phase behaviour of water is a nonlinear function of temperature: by analogy with the Maxwell distribution of speeds in gases, the probability that an individual molecule will have a speed much in excess of the average increases with temperature: evaporation occurs more readily at higher temperatures. Figure 3.17 shows the effect of temperature and pressure on evaporation rate.

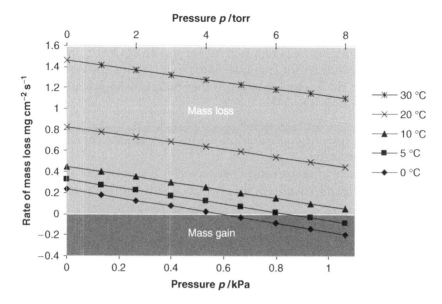

Figure 3.17   Plot of mass loss for water due to evaporation as a function of pressure. If the rate is sufficiently low, the pressure of the environment may be somewhat below the equilibrium vapour pressure of water without causing significant dehydration over a finite period. Calculated data originally courtesy of Brad Thiel, SUNY Albany

This kinetic behaviour has important implications in the case of controlling water in and around specimens in the VP-ESEM chamber. Typical operating temperatures for hydrated specimens tend to be around 2–6 °C, where the rate of moisture loss is really quite low. It is therefore acceptable to employ pressures somewhat below the equilibrium vapour pressure given by the modified SVP curve in Figure 3.16: real specimens can usually withstand slowly dehydrating conditions for a finite period of time (tens of minutes if the temperature is, say, a degree or so).

3.4.4 Practicalities of Stabilising Hydrated Specimens

Section 3.4.2 outlined the conditions (temperature and water vapour pressure) that are needed to maintain an aqueous phase in its fully hydrated state. It is worth emphasising that, despite deviating to conditions well below 100 % relative humidity, the specimen itself is *not* under dehydrating conditions (Stokes, 2003). Indeed, if a specimen is held at a water vapour pressure higher than its equilibrium pressure, water will condense onto its surface, rather like the situation shown in Figure 3.14. This has been reported by Tai and Tang (2001) and verified by other workers (see, for example, Muscariello *et al.*, 2005).

Meanwhile, there are a couple of other practical considerations that should be mentioned here:

1. the risk of specimen dehydration whilst pumping the chamber to the required pressure;
2. the difference in temperature between the specimen and the source of vapour used to maintain the correct conditions above the (cooled) specimen.

Having specified equilibrium conditions for a given specimen, we must first get from a chamber that contains air at atmospheric pressure to one that contains the prescribed pressure of water vapour with the specimen at an appropriate temperature. One way to do this is through 'purge–flood' cycles, in which air is systematically replaced with water vapour. Work by Cameron and Donald (1994) graphically showed the importance of using the correct parameters to minimise any water loss during pumpdown. Their work led to a widely implemented purge–flood cycle, described below.

The procedure recommended by Cameron and Donald for a specimen temperature of $3\,°C$ is to pump to a pressure $p = 731.5\,Pa$ (5.5 torr), allow water vapour into the chamber until the pressure rises to $p = 1.3\,kPa$ (9.6 torr) and repeat eight times, finishing at $p = 731.5\,Pa$ (5.5 torr). Note, of course, that the specified values are for conditions of 100 % RH. For a specimen having an equilibrium vapour pressure 75 % RH, the corresponding lower-limit pressure for this temperature would be $p = 545\,Pa$ (4.1 torr).

However, some specimens may still be vulnerable to the evaporation of water before the purge–flood cycle begins. A very simple method to deal with this is to place small drops of water on a noncooled area near the specimen. When the humidity in the chamber falls, these droplets will sacrificially evaporate, having higher vapour pressure than the cooled specimen, giving a much-needed burst of vapour. Another method involves surrounding the specimen with a medium such as agar gel, having a higher vapour pressure, which will similarly evaporate in preference to the specimen itself (Nedela, 2007).

Turning our attention to the vapour being delivered to the specimen chamber, consider a source of vapour held at room temperature T_r, a specimen cooled to $T_s = 3\,^{\circ}C$ and water vapour pressure $p_0 = 545\,Pa$ (4.1 torr). There will be a slight imbalance between the number of vapour molecules arriving at the surface compared to those leaving (a greater number arriving), sufficient to cause a small downward shift Δp in the required vapour pressure.

If this shift is not taken into account and the operating pressure lowered accordingly, water may condense onto the specimen from the warmer source vapour. Equation (3.12) (after Cameron and Donald, 1994), gives an expression to correct for the difference in temperature when plotting the SVP curve for water:

$$p_{0(\text{corrected})} = \left(\frac{T_s}{T_r}\right)^{1/2} \cdot p_0(T_s) \tag{3.12}$$

where $p_0(T_s)$ is simply the saturated vapour pressure p_0 for pure water as a function of specimen temperature T_s (as per Figure 3.28 and Table 3.2) and T_r is room temperature. Note that T is in Kelvin.

Using Equation (3.13) below, we can quickly deduce the impact that this has on the value of p_0 for a given set of conditions, by determining the difference in pressure Δp, as follows:

$$\Delta p = p_0 - p_{0(\text{corrected})} \tag{3.13}$$

For example, if $p_0 = 545\,Pa$ (4.1 torr), $T_s = 3\,^{\circ}C$ (276 K) and $T_r = 20\,^{\circ}C$ (293 K), Equation (3.12) gives $p_{0(\text{corrected})} = 529\,Pa$ (3.98 torr), and so Equation (3.13) gives a value for Δp of 16 Pa (0.12 torr).

Finally, combining Equations (3.12) and (3.13) gives an expression for the equilibrium vapour pressure p_{eq} of a specific aqueous phase that takes into account both the above correction and the specimen's own equilibrium vapour pressure (Equation (3.14)).

$$p_{eq} = a_w p_0 \left[\frac{T_s}{T_r}\right]^{1/2} \tag{3.14}$$

Recall that p_0 is the value of p for 100 % RH and a_w is the water activity of the aqueous phase (see Section 3.4.2).

Finally, further lowering of the chamber pressure can be accommodated, due to kinetic factors (discussed in Section 3.4.3). In fact, for some specimens, it is possible to work at a relative humidity of just 50 % (i.e. relative to pure water) for about 20–30 minutes per specimen. At $3\,^{\circ}C$, this means a pressure $p_{eq} \sim 305\,Pa$ ($\sim$2.3 torr).

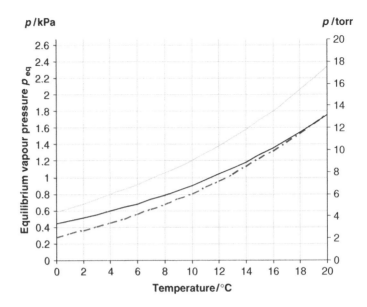

Figure 3.18 Plot of equilibrium vapour pressure p_{eq} (solid black line) assuming a specimen with water activity $a_w = 0.75$ (equivalent to an equilibrium relative humidity RH = 75 %), corrected for a water vapour source at $T = 20\,°C$ (293 K) and specimen temperature $T_s = 3\,°C$ (276 K). The dashed line is an approximate indicator of permissible short -term (meta)stability arising from kinetic factors and the grey line is the saturated vapour pressure for pure water

Putting together these various thermodynamic and kinetic factors we can develop a further modified phase diagram that takes account of the specimen's equilibrium vapour pressure and rate of water loss. An example is shown in Figure 3.18 for an equilibrium relative humidity RH = 75 %. The dashed line gives an indication (based on empirical observations) as to the short-term reduction in pressure that is afforded by the slow rate of water loss at lower temperatures. Aside from avoiding any unnecessary condensation of water on the specimen, another effect of reducing the pressure to the values indicated is that it helps to reduce scattering of the primary beam. This is a subject that we will explore in detail in Chapter 4.

REFERENCES

Boeuf, J.P. and Marode, E. (1982). A Monte Carlo analysis of an electron swarm in a non-uniform field: the cathode region of a glow discharge in helium. *J. Phys. D: Appl. Phys.*, **15**, 2169–2187.

Cameron, R.E. and Donald, A.M. (1994). Minimising Sample Evaporation in the Environmental Scanning Electron Microscope. *J. Microsc.*, **173**(3), 227–237.

Danilatos, G.D. (1986). Cathodoluminescence and gaseous scintillation in the environmental SEM. *Scanning*, **8**, 279–284.

Danilatos, G. (1990a). Mechanisms of Detection and Imaging in the ESEM. *J. Microsc.*, **160**(pt. 1), 9–19.

Danilatos, G.D. (1990b). Theory of the Gaseous Detector Device in the Environmental Scanning Electron Microscope. *Adv. Electron.Elect. Phys.*, **78**.

Durkin, R. and Shah, J.S. (1993). Amplification and Noise in High Pressure Scanning Electron Microscopy. *J. Microsc.*, **169**, 33–51.

Ellis, R.J. (2001). Macromolecular Crowding: Obvious But Underappreciated. *Trends Biochem. Sci.*, **26**(10), 597–604.

Farley, A.N., Beckett, A. and Shah, J.S. (1990). *Comparison of Beam Damage of Hydrated Biological Specimens in High-Pressure Scanning Electron Microscopy and Low-Temperature Scanning Electron Microscopy.* Proc. XIIth International Congress for Electron Microscopy, San Francisco Press.

Farley, A.N. and Shah, J.S. (1991). High-Pressure Scanning Electron-Microscopy of Insulating Materials – A New Approach. *J. Microsc.–Oxford*, **164**, 107–126.

Fletcher, A., Thiel, B. and Donald, A. (1997). Amplification measurements of Potential Imaging Gases in Environmental SEM. *J. Phys. D: Appl. Phys.*, **30**, 2249–2257.

Goldstein, J., Newbury D, Joy, D, Lyman, C, Echlin, P, Lifshin, E, Sawyer, L and Michael, J (2003). *Scanning Electron Microscopy and X-Ray Microanalysis*, third edition, Plenum.

Lide, D.R. (Ed.) (1991). *CRC Handbook of Chemistry and Physics*, CRC Press.

Meredith, P., Donald, A.M. and Thiel, B. (1996). Electron–gas interactions in the environmental scanning electron microscope's gaseous detector. *Scanning*, **18**(7), 467–473.

Mohan, A., Khanna, N., Hwu, J. and Joy, D.C. (1998). Secondary electron imaging in the variable pressure scanning electron microscope. *Scanning*, **20**, 436–441.

Moncrieff, D.A., Robinson, V.N.E. and Harris, L.B. (1978). Gas neutralisation of insulating surfaces in the SEM by gas ionisation. *J. Phys. D: Appl. Phys.*, **11**, 2315–2325.

Morgan, S.W. and Phillips, M.R. (2005). Transient analysis of gaseous electron–ion recombination in the environmental scanning electron microscope. *J. Microsc.*, **221**(pt 3), 183–202.

Morgan, S.W. and Phillips, M.R. (2006). Gaseous scintillation detection and amplification in variable pressure scanning electron microscopy. *J. Appl. Phys.*, **100**(7), Article no. 074910.

Muscariello, L., Rosso, F., Marino, G., Giordano, A., Barbarisi, M., Cafiero, G. and Barbarisi, A. (2005). A critical review of ESEM applications in the biological field. *J. Cellular Phys.*, **205**, 328–334.

Nedela, V. (2007). Methods for additive hydration allowing observation of fully hydrated state of wet samples in environmental SEM. *Microsc. Res. Techn.*, **70**(2), 95–100.

Phillips, M.R., Toth, M. and Drouin, D. (1999). Depletion layer imaging using a gaseous secondary electron detector in an environmental scanning electron microscope. *Appl. Phys. Lett.*, **75**(1), 76–78.

Reimer, L. (1985). *Scanning Electron Microscopy. Physics of Image Formation and Microanalysis*, Springer-Verlag.

Stokes, D.J. (2003). Recent advances in electron imaging, image interpretation and applications: environmental scanning electron microscopy, *Phil. Trans. Roy. Soc. Lond. Series A-Math. Phys. Eng. Sci.*, **361**(1813), 2771–2787.

Tabor, D. (1991). *Gases, Liquids and Solids, and Other States of Matter*, Cambridge University Press.

Tai, S.S.W. and Tang, X.M. (2001). Manipulating Biological Samples for Environmental Scanning Electron Microscopy Observation. *Scanning*, **23**, 267–272.

Thiel, B.L. (2004). Master curves for gas amplification in low vacuum and environmental scanning electron microscopy. *Ultramicrosc.*, **99**(1), 35–47.

Thiel, B.L., Bache, I.C., Fletcher, A.L., Meredith, P. and Donald, A.M. (1997). An Improved Model for Gaseous Amplification in the Environmental SEM. *J. Microsc.*, **187**(pt. 3), 143–157.

Toth, M. and Phillips, M.R. (2000). The role of induced contrast on images obtained using the environmental scanning electron microscope. *Scanning*, **22**, 370–379.

von Engel, A. (1965). *Ionized Gases*, Clarendon Press.

4

Imaging and Analysis in VP-ESEM: The Influence of a Gas

4.1 INTRODUCTION

Now that we have seen why and how a gas is used in VP-ESEM, the major focus of this next chapter is to take a quantitative look at what happens when different gases are used and to consider how the concentrations of gas atoms or molecules influence the mean free paths, trajectories and spatial distributions of primary electrons. All of these factors have a significant effect on the useful primary electron beam current and hence the quality of results.

We begin with some important notes about the basis of the calculations and simplifying assumptions used in this chapter, before moving on to make some quantifications that will help us to determine the trends in primary electron behaviour in a gaseous environment. Whilst many of these results are theoretical, they allow us to explore a wide range of circumstances and are in broad agreement with the properties that the VP-ESEM user is likely to encounter in practice.

Throughout this chapter there is some discussion of the measures employed to help minimise primary electron scattering due to the influence of a gas, leading to a strategy for more effective imaging, particularly at lower voltages, plus improved X-ray microanalysis.

Principles and Practice of Variable Pressure/Environmental Scanning Electron Microscopy (VP-ESEM)
D. J. Stokes
© 2008 John Wiley & Sons, Ltd

4.2 BACKGROUND TO THEORETICAL CALCULATIONS

4.2.1 Calculating the Mean Free Paths of Primary Electrons

Recall that in Chapter 2 we looked at the determination of primary electron scattering cross-sections σ and mean free paths λ. Now, to be strictly accurate, we should use the *total* scattering cross-section σ_T, which depends on numerous factors, both elastic and inelastic. But, for simplicity, we will use only the Rutherford elastic scattering cross-section σ_e, defined in Chapter 2, Section 2.4.1.1, to calculate primary electron scattering in a gas, and hence the elastic mean free path λ_e. This means that we will neglect the contribution to the total scattering cross-section of inelastic processes such as excitation and ionisation, and so the numerical values we arrive at are to be treated as an indication only. On the plus side, the angles through which primary electrons are deflected in elastic events is roughly half an order of magnitude higher than for inelastic events ($5°$ or more, compared to $0.1°$, respectively), and the important question is: *where* do the primary electrons end up? If we are interested in knowing which electrons are in the focused probe and which are in the skirt, then we could argue that the elastic scattering process is of greater significance. If, however, we are interested in knowing about the *energies* of inelastically scattered primary electrons, which lose energy but are scattered through very small angles and may therefore remain in the focused probe, then that is a different matter, presently beyond the scope of this book.

There are two further assumptions in the calculations that should be stated. One is that primary electrons elastically scattered through an angle $1°$ or less are considered to remain in the focused probe (a similar approach is used in Newbury, 2002; Goldstein *et al.*, 2003 and Tang and Joy, 2005). The other is that only single or oligo-scattering takes place (see Chapter 3, Section 3.2.2), so that we do not have to consider the consequences of multiple scattering.

4.2.2 Calculating Pressure-Dependent Variables

Ultimately, in order to calculate mean free paths of electrons in a gas, we need additional information about how the density of the gas varies as a function of pressure. An alternative form of the ideal gas law is very useful for our purposes, and conveniently gives the density ρ as:

$$\rho = pM/RT \qquad (4.1)$$

where p = pressure (Pa), M = molar mass (g/mol), R = universal gas constant ($8314472 \, cm^3 \cdot Pa \cdot K^{-1} \cdot mol^{-1}$) and T = temperature (Kelvin).

Note that, herein, the values of molar mass used in any calculations involving nitrogen and oxygen reflect their diatomic nature, i.e. they naturally exist as the molecular entities N_2 and O_2 and have molar masses approximately 28 g/mol and 32 g/mol, respectively. This has an impact on elastic mean free path calculations compared to using values for the monatomic molar masses (N ~14 g/mol, O ~16 g/mol).

In Chapter 2, Equation (2.7) gave the relationship between scattering probability, atomic mass and density, and hence the mean free path λ of electrons in a given material, with the scattering probability being deduced from Equation (2.6). Equation (4.1) can thus be used to insert the appropriate values for the pressure-dependent density into Equation (2.7), enabling us to compare elastic mean free paths as a function of pressure as well as other variables such as atomic number and primary beam energy.

Combining Equations (2.7) and (4.1) we have an expression for calculating the gas pressure-dependent elastic mean free path of primary electrons λ_e (Equation (4.2)):

$$\lambda_e = \frac{ART}{N_0 \sigma \rho M} \qquad (4.2)$$

Note that, in keeping with previous practice, the units of λ_e are centimetres.

4.2.3 Estimating the 'Useful' Primary Electron Current

In Chapter 3 we made a distinction between working distance WD and the anode–specimen distance d (Section 3.3.2.1). Recall that the working distance WD is given by the distance between the end of the objective lens polepiece and the specimen. If an on-axis anode is being employed to detect the cascaded secondary electron signal, also serving as the final pressure-limiting (differential) aperture, then d is the distance that primary electrons travel through the gas on the way to the specimen, measured from the end of the aperture. This distance is the gas path

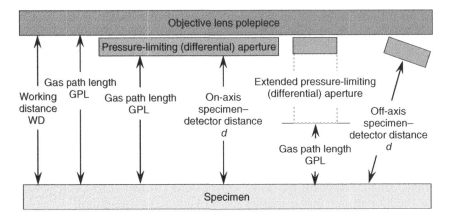

Figure 4.1 Schematic diagram to define working distance *WD*, specimen–anode distance *d* and gas path length GPL for different configurations. The gas path length is measured as the distance through which primary electrons travel in a gas on the way to the specimen. Depending on the specific configuration, GPL may correspond to the value *WD* or *d*, or may differ from these values as a result of an extended pressure-limiting (differential) aperture

length, GPL.[1] However, different geometries can be employed so that the specimen–anode distance *d* and the gas path length GPL do not have the same numerical value. These ideas are exemplified in Figure 4.1, and we will come back to this in Section 4.4.

From this point on then, when referring to gas path length GPL, this strictly means the distance between any final aperture and the specimen.

Now, if we consider the number of collisions *m* encountered by a molecule with mean free path λ travelling a distance equivalent to the gas path length GPL, we can deduce the average number of scattering events, according to:

$$m = \text{GPL}/\lambda \tag{4.3}$$

Based on this, it is possible to determine, to a first approximation, the fraction f_p of electrons that are minimally scattered and can therefore be thought of as remaining in the central focused probe. This also gives us a feel for the way in which the useful signal-forming primary beam current is likely to change with parameters such as beam energy, gas path length and atomic number of the gas. If we assume that the scattering

---

[1] For primary electrons, this is sometimes referred to specifically as the beam gas path length, BGPL.

follows a Poisson distribution, then it can be shown that the fraction of unscattered electrons is described by Equation (4.4):

$$f_p = e^{-m} \qquad (4.4)$$

Expressing Equation (4.3) as a percentage, we have:

$$\text{Percentage of electrons in focused probe}/\% = f_p \cdot 100 \qquad (4.5)$$

If the average number of scattering events is $m = 1$, for example, then $f_p = exp(-1) = 0.37$, so 37 % of the beam remains in the focused probe. But if m is outside the single or oligo-scattering boundary conditions mentioned earlier, say $m = 4$, then $exp(-4) = 0.018$, or 1.8 % of electrons are unscattered. We would not therefore expect to produce an image-forming probe under these circumstances. Note that, for the purposes of VP-ESEM, the upper limit of m for oligo-scattering is arbitrarily defined as $m = 3$ (Danilatos, 1994b). This effectively says that we can still produce an image-forming probe with only 5 % of the primary electrons remaining.

So if we now combine Equations (4.3) and (4.4), we have an expression (Equation (4.6)) that enables us to calculate f_p for a variety of gas path lengths and mean free paths:

$$f_p = e^{(-GPL/\lambda)} \qquad (4.6)$$

Accordingly, the fraction f_s of electrons that collide with gas molecules to end up in the skirt, is given by:

$$f_s = 1 - e^{(-GPL/\lambda)} \qquad (4.7)$$

Finally, the useful primary beam current i_p can be quantified by taking the product of the initial beam current i_0 and the fraction of electrons that remain in the probe f_p, as deduced from Equation (4.6), i.e.:

$$i_p = f_p \cdot i_0 \qquad (4.8)$$

We now have almost everything we need to carry out some quantitative analyses of primary electron behaviour in different gases. All that remains is to choose some representative gases for which data are readily available for all the parameters needed. For a wide range of properties, the gases helium, nitrogen, oxygen and argon will serve our purposes well, and all are gases that can be used in VP-ESEM. Note that although water vapour and air are the most commonly used gases, it becomes yet more subjective to define appropriate analytical parameters for the scattering cross-sections of compounds and mixtures. Given that we have already made a number of simplifying assumptions and it has been noted that the numerical values are to be taken only as a guide, it will be sufficient to note that the trends observed for nitrogen and oxygen will be comparable to those we would expect for water and air. The interested reader can find a few articles in the literature involving primary electron scattering of gases in VP-ESEM, for example: Danilatos (1988), Mathieu (1999), Goldstein *et al.* (2003), Kadoun *et al.* (2003) and Thiel *et al.* (2006). Different approaches are used for calculating elastic cross-sections and, again, care should be taken in interpreting such information, since experimental verification is extremely difficult to obtain.

4.3 WHICH GAS?

4.3.1 Introduction

As mentioned before, a number of different gases have been tried in VP-ESEM, with the most commonly used being air and water vapour. Others include argon, nitrogen, helium, carbon dioxide and nitrous oxide. At a fundamental level the physical properties of a gas, such as atomic number, excitation probability, ionisation energy and scattering cross-section, determine whether it makes a suitable gas for a given experiment in VP-ESEM. The effects of these properties are explored in the following sections and quantified wherever possible. To begin, we discuss some of the reasons why we would want to consider using alternative gases in the first place.

4.3.2 Usefulness of the Gas – Experimental Conditions

The purpose of this section is to briefly highlight the important properties of some potential imaging gases in relation to the type of observation or experiment required in VP-ESEM.

Air and water vapour are frequently used as they are inexpensive, readily available and easy to handle. Water vapour is a special case in VP-ESEM, for several reasons: it is relatively easy to ionise and hence generate an electron amplification cascade and, when used in conjunction with specimen temperature, can be used to control the hydration state of a specimen. This was discussed in detail in Chapter 3, Section 3.4. The signal detection mechanism is also an important criterion for choosing an appropriate imaging gas. For example, certain properties of gases other than water vapour and air may lend themselves better to the collection of the gas luminescence signal. This will be discussed further in Section 4.3.3.

Meanwhile, for specimens that oxidise easily, argon may be chosen for its inertness. For experiments at low temperatures, gases such as helium, nitrogen and nitrous oxide may be selected for their high vapour pressures and hence resistance to precipitating as a liquid or solid when cold. Gases such as ethanol or xenon difluoride may be chosen to specifically react with a given substrate, or perhaps an inert gas such as argon may be needed to prevent certain reactions, particularly at higher temperatures. Others may be used simply to provide the necessary vapour pressure of an appropriate gas to stabilise the specimen (e.g. water vapour, carbon dioxide). In Chapter 6 we will cover some of the applications that involve these different kinds of gases in VP-ESEM.

4.3.3 Ionisation and Excitation for Different Gases

Ionisation is clearly an important parameter when considering the amplification of secondary electrons via the gas cascade mechanism and, in order to make some comparisons, Table 4.1 shows the first ionisation potentials[2] for a few different gases that can be used for imaging in VP-ESEM.

From Table 4.1 we see that water vapour is the most easily ionised of the gases listed, whilst helium is the most difficult. This is consistent with the trends in gaseous secondary electron signal gain observed in VP-ESEM (see, for example, Danilatos, 1988; Fletcher *et al.*, 1999).

Table 4.1 Ionisation potentials for a range of gases used in VP-ESEM (source: NIST)

Gas	First ionisation potential/eV
Water vapour	12.6
Oxygen	13.6
Nitrogen	14.5
Argon	15.8
Helium	24.6

[2] That is, the energy to remove the lowest-energy electron from the neutral atom or molecule. The energy needed to ionise subsequent electrons will be increasingly higher.

Meanwhile, as mentioned in Chapter 3, Section 3.3.3.1, there is the possibility of exploiting subtle differences between the ionisation-related secondary electron cascade signal compared to the gas luminescence photon signal, based on differences between the energy needed to excite a given atomic orbital transition versus the energy needed to cause ionisation and form a gaseous ion. These excitation energies are, by definition, lower than the ionisation potential, and so the onset of photon production from excitation–relaxation events is likely to occur before that related to ionisation and electron–ion recombination – perhaps even in preference to it, under certain conditions of anode bias, etc.

Data on the appropriate excitation energies, probabilities and scintillation efficiencies would be helpful at this point, but would require spectroscopic studies for the specific gases, pressures and electric field strengths employed in VP-ESEM that are not presently forthcoming in the literature.

However, the work of Morgan and Phillips (2006) does throw some light on the situation. It is found that the gas luminescence signal gain is greatest for argon, followed by nitrogen and then water vapour: the opposite of the trend for secondary electron gain (although the gain falls off rapidly with increasing pressure). In general, the amplification of photon signals is found to match closely the pattern of amplified electron signals as the grid or anode biases, respectively, are changed. Likewise, the characteristics are similar with changes in working distance. However, when the chamber gas pressure is varied for argon and nitrogen, the photon signal is initially greater than the gas-amplified electron signal, but tails off beyond about 266 Pa (2 torr) while the electron signal continues to increase. For water vapour, a sharp drop-off in photon signal is seen beyond ~133 Pa (1 torr). That said, the signal-to-background ratio was reported to be significantly better for the photon signal, confirming the more dominant contribution to the total signal by primary and backscattered electrons to the gas ionisation amplification process, as was indicated in Figure 3.7.

Now, whilst the information so far may be helpful in determining certain properties of an imaging gas, namely its image-forming capabilities, it does not tell us the whole story. The gas also has a profound effect on the scattering of primary electrons before the signal is even formed. Hence, we must also consider the effects of the imaging gas on primary electron mean free paths, on the ratio of the useful electron current to that scattered into the delocalised beam skirt and the extent of the skirt. This will now occupy us for the remainder of the chapter.

4.3.4 Scattering of the Primary Electron Beam in Different Gases

4.3.4.1 The Influence of Atomic Number on the Elastic Mean Free Path

In accordance with the explanations given in Section 4.2, we are now equipped to get a more quantitative feel for primary electron scattering in the oligo-scattering regime encountered in the gaseous environment of VP-ESEM. Figure 4.2 shows a plot of the elastic mean free path λ_e as a function of atomic number for a range of primary electron beam energies, $5\,keV < E_0 < 30\,keV$. Values are given for a constant gas pressure $p = 100$ Pa (0.75 torr) and the units are centimetres.

In Chapter 2 (Section 2.4.1.1) we saw that the probability for elastic scattering increases with increasing atomic number Z and with decreasing beam energy E_0. Figure 4.2 confirms this trend, showing that gases with lower atomic numbers produce a smaller amount of elastic scattering and hence result in longer elastic mean free paths λ_e, and that λ_e is longest for higher beam energies. Specifically, we see that for primary beam energy $E_0 = 30\,keV$, λ_e is approximately 40 cm for helium, while

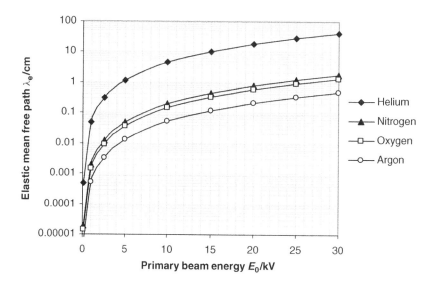

Figure 4.2 Log-linear plot of primary electron mean free paths as a function of atomic number for a range of primary beam energies. The data points are for helium ($Z = 2$), nitrogen ($Z = 7$), oxygen ($Z = 8$) and argon ($Z = 18$). Pressure $p = 100$ Pa (0.75 torr)

for nitrogen, oxygen and argon, λ_e goes down from about 2 cm to 1.5 cm to 5 mm, respectively. Meanwhile, for $E_0 = 10\,keV$, the values decrease from about 1 cm for helium to 2 mm for nitrogen, 1.5 mm for oxygen and 0.5 mm for argon.

If we take typical gas path lengths in VP-ESEM, as defined in Section 4.2.3, Figure 4.1, to be anywhere between 1 mm and 15 mm, this immediately gives an idea of how likely it is that primary electrons will be scattered to a significant degree. For example, for nitrogen gas at a pressure $p = 100$ Pa, primary beam energy $E_0 = 10\,keV$ and a working distance $WD = 1\,mm$, the primary electron beam, with elastic mean free path $\lambda_e \sim 2\,mm$, will suffer relatively few elastic scattering events. However, to achieve a similar effect using argon, the primary beam energy would have to be increased to $E_0 \approx 20\,keV$.

Imaging at long gas path lengths (e.g. 10–15 mm) is quite feasible, but gets much more difficult for lower energies, where the primary electron mean free path becomes somewhat shorter than the gas path length. In addition, when the gas pressure is rather higher than demonstrated here, scattering can become excessive. This will be shown in Section 4.4.

4.3.4.2 Effect of Atomic Number on the Radius of the Primary Beam Skirt

Now let us consider the electrons that are scattered into the diffuse skirt region around the central focused beam. An important implication of the results in Figure 4.2 is that we might expect the primary beam skirt to be smaller for a light gas such as helium ($Z = 2$) compared to other gases. Conversely argon, with its higher atomic number ($Z = 18$) and much stronger effect on electron scattering, might be expected to result in a more diffuse skirt (i.e. more widely spread), with nitrogen and oxygen ($Z = 7$ and 8, respectively) being intermediate. Indeed, the primary electron skirt becomes a very important factor when carrying out X-ray microanalysis (we will return to this point in Section 4.6). Next, we check if these assumptions are correct, and calculate some values for the extent of the skirt for different gases.

According to Danilatos (1988), the radius of the skirt r_s can be described analytically by the following expression:

$$r_s = (364Z/E_0)(p/T)^{1/2}GPL^{3/2} \qquad (4.9)$$

Equation (4.9) is plotted in Figure 4.3 to show the variation in skirt radius as the atomic number increases, as a function of primary beam

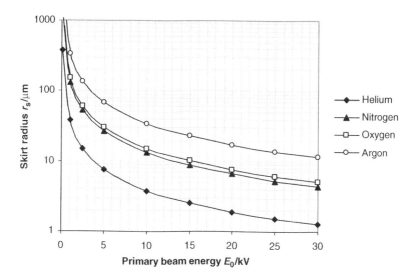

Figure 4.3 Log-linear plot of skirt radius r_s as a function of primary beam energy for a range of gases having atomic numbers $Z = 2$ (helium), $Z = 7$ (nitrogen), $Z = 8$ (oxygen) and $Z = 18$ (argon). Thickness of the gas layer (gas path length) $= 2$ mm. The temperature is assumed to be $T = 293$ K ($20\,^\circ$C) and pressure $p = 100$ Pa (0.75 torr)

energy, assuming that the primary electron beam travels through a thickness of 2 mm of gas at a pressure of 100 Pa ($\sim$0.75 torr). We can see that, indeed, the skirt radius goes up quite markedly with increasing atomic number but decreases with increasing primary beam energy.

Figure 4.3 thus gives us a feel for the way the skirt radius changes with the atomic number of the gas, as a function of primary beam energy E_0. As we can see, the scattering radius is directly proportional to atomic number. Increasing the primary beam energy significantly reduces the distance to which primary electrons are scattered. The plot shows that the skirt radii for nitrogen and oxygen are quite similar, ranging from around 30 µm at 5 keV down to about 4 µm at 30 keV for this set of conditions (GPL = 2 mm, pressure $p = 100$ Pa, 0.75 torr). The primary beam skirt radius for helium is just over a micron at 30 keV, while for argon the value goes up to more than 10 µm.

In addition to Equation (4.9), Monte Carlo simulations can be used to predict the size of the primary beam skirt and also the overall shape of the beam profile (Mathieu, 1999; Kadoun et al., 2003; Tang and Joy, 2005).

Some experimental work has been carried out in order to test these theoretical predictions. One method is to capture and measure the intensity of electrons falling on a YAG scintillator crystal, via the

Figure 4.4 Experimental verification of the shape of the primary electron beam skirt in VP-ESEM (three-dimensional reconstruction). Notice the finely focused probe in the centre that produces the information-carrying signal and defines resolution. Imaging gas is water vapour with pressure $p = 266$ Pa (2 torr), primary beam energy $E_0 = 20$ keV and gas path length GPL $= 3$ mm. Courtesy of Brad Thiel, SUNY, Albany

emitted photons, a lens system and a CCD camera (Thiel *et al.*, 2000). Measurements were made for helium, water vapour, nitrogen and nitrous oxide, all showing very similar profiles. The intensity of the skirt, and hence the degree of scattering, was found to increase in the order: helium < water vapour < nitrogen < nitrous oxide. A three-dimensional reconstruction of the two-dimensional data collected for water vapour is shown in Figure 4.4.

Meanwhile, several other experiments have been carried out, for example by Wight *et al.* (1997), Gillen *et al.* (1998) and Wight and Zeissler (2000), involving the effects of electron beam damage in self-assembled decanethiol monolayers, which can be used to construct a beam-skirt profile in conjunction with three-dimensional secondary ion mass spectrometry (SIMS). Reimer (1985) has shown that gas luminescence can be used to directly visualise the nature of the scattering of primary electrons in gases (demonstrating the forward-peaked behaviour of scattering in nitrogen compared to argon, for example). Another way to infer the extent of the skirt electrons is by detecting spurious X-ray signals (see Section 4.6) generated by elements remote from the beam impact point (Goldstein *et al.*, 2003).

4.3.4.3 Influence of Atomic Number on the Useful Primary Electron Beam Current

Given that electrons falling into the skirt do not contribute to the useful primary electron beam current, Figure 4.3 implies that there is a large reduction in beam current when the primary electron beam energy E_0 is low. As noted in Chapter 3, Section 3.2.2, the loss of primary electrons from the focused part of the electron beam causes a loss of primary beam current. Whilst this does not generally affect resolution (the same feature sizes can still be seen), it does decrease the signal-to-noise ratio, so images appear more 'grainy' and the contrast goes down, making it more difficult to distinguish adjacent features having similar electron emission characteristics.

One way to improve the situation, especially if low-voltage imaging or X-ray microanalysis are required, is to reduce the distance that the primary electrons have to travel in the gas, i.e. the gas path length GPL. We will look in detail at the effects of gas path length in Section 4.4 and again in Section 4.6. For the moment, we will try to quantify the effect that atomic number has on the useful primary beam current.

Figure 4.5 shows the fractions of electrons remaining in the focused probe as a percentage, calculated from Equation (4.6), again assuming a gas path length GPL = 2 mm and pressure $p = 100$ Pa (0.75 torr). We can immediately infer that there is indeed a direct correlation: the smaller the size of the primary beam skirt radius r_s, the higher the likelihood of electrons remaining in the probe to form a useful current. For example, we saw in Figure 4.3 that the scattering of primary electrons in argon forms a sizable beam skirt, so we may expect that a large fraction of primary electrons are lost in this way. From Figure 4.5 we see that, for example, 40 % of the original primary electrons remain in the probe to form the useful beam current at 20 keV in argon. Conversely helium, whose skirt radius is estimated to be much smaller, loses hardly any primary electrons from the probe under these conditions. Keep in mind that the theoretical arbitrary minimum requirement for an image-forming probe is to have 5 % of the original primary electrons remaining in the probe (Section 4.2.3).

The results shown so far then, tend to indicate that helium is a good candidate for minimising beam–gas interactions, and a few relevant

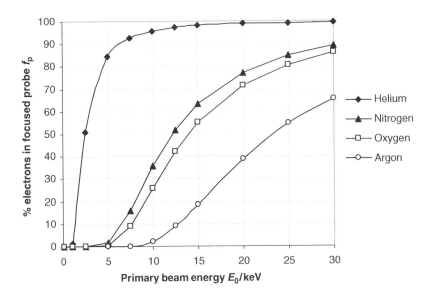

Figure 4.5 Plot of the percentage of primary electrons remaining in the focused probe to form a useful beam current, as a function of primary electron beam energy E_0 for a range of gases with atomic numbers $Z = 2$ (helium), $Z = 7$ (nitrogen), $Z = 8$ (oxygen) and $Z = 18$ (argon). Thickness of gas layer (gas path length) $= 2\,\mathrm{mm}$, pressure $p = 100\,\mathrm{Pa}$

studies have been reported in the literature (Stowe and Robinson, 1998; Kadoun *et al.*, 2003). Correspondingly, however, it is the most difficult gas to ionise and the small size of helium atoms makes this gas notoriously difficult to handle with a typical vacuum pump. In fact, prolonged use of helium (more than a few hours at time) is definitely not recommended. Besides, there are many other good reasons to use alternative gases, as mentioned in Section 4.3.2.

4.4 EXPLORING THE GAS PATH LENGTH

4.4.1 Introduction

So far we have seen that the elastic mean free path of primary electrons decreases with increasing atomic number Z and that the effects of elastic scattering get worse as the primary beam energy decreases. In Section 4.3.4.1 we noted that if the elastic mean free path λ_e for a given pressure p is comparable to or longer than the gas path length GPL, electron scattering is minimised. Now let us think a little about the gas path length: in particular, how it affects the primary electron beam when the gas path length is either longer or shorter than the elastic mean free path, and how the geometry of the system can be optimised.

4.4.2 Influence of GPL on the Skirt Radius

Recall that the primary electron beam skirt radius r_s involves a dependence on gas path length (see Equation (4.8)). Figures 4.6 to 4.8 therefore show how r_s changes for gases of different atomic number ($Z = 2$, 7 and 18, respectively) for a range of gas path lengths and primary beam energies at a pressure $p = 100$ Pa (0.75 torr). Note that the results for each atomic number are plotted separately for clarity, and that the y-axis scale for the helium plot is a factor of 10 lower than for the others. Data for oxygen have been omitted as they are within 9 % (higher) of the values for nitrogen and are therefore very similar on this scale.

The general trends can be seen in all cases: there is a sharp rise in the radius of the skirt as the primary beam energy decreases below about 5 keV, and almost two orders of magnitude difference between radii for gas path lengths GPL $= 1$ mm compared to 15 mm, helping to reinforce the important influence of the gas path length on primary electron scattering.

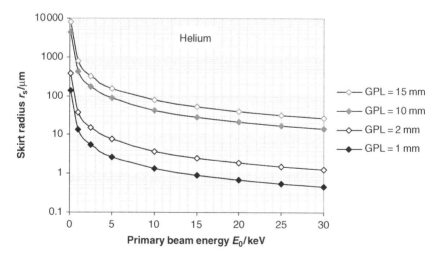

Figure 4.6 Plot of skirt radius as a function of gas path length and beam energy E_0 for primary electrons in helium gas. Pressure $p = 100$ Pa (0.75 torr)

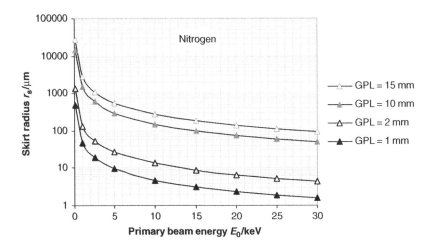

Figure 4.7   Plot of skirt radius as a function of gas path length and beam energy E_0 for primary electrons in nitrogen gas. Pressure $p = 100$ Pa (0.75 torr)

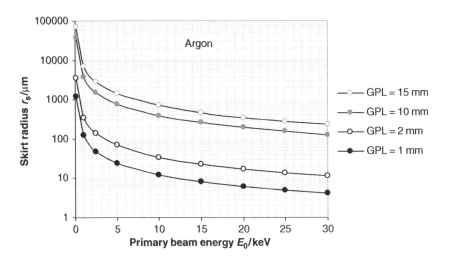

Figure 4.8   Plot of skirt radius as a function of gas path length and beam energy E_0 for primary electrons in argon gas. Pressure $p = 100$ Pa (0.75 torr)

4.4.3 Gas Path Length and Useful Primary Electron Beam Current

Figure 4.9 compares the percentage of electrons remaining in the focused probe as a function of gas path length GPL in the range $0 \leq$ GPL $\leq$

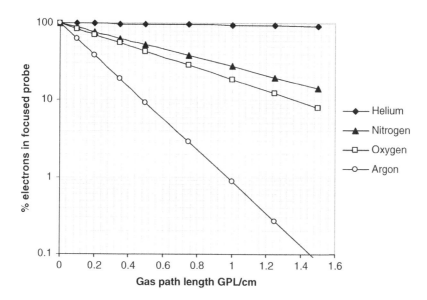

Figure 4.9 Log-linear plot to show the percentage of electrons remaining in the focused probe to form the useful primary current, as a function of atomic number of the gas for a primary beam energy $E_0 = 20\,\text{keV}$. Pressure $p = 100$ Pa

1.5 cm, for different gases and a primary beam energy $E_0 = 15\,\text{keV}$. With the exception of helium, the percentage of electrons remaining falls sharply as the gas path length is increased. This brings with it a reduction in electron beam current. For argon, it would seem advisable to maintain a gas path length of a millimetre or two. For nitrogen and oxygen, gas path lengths of the order of a few millimetres may ensure an adequate proportion of electrons for imaging if necessary, although shorter gas path lengths would clearly be more beneficial.

4.4.4 Constraints on Reducing the Gas Path Length

If we assume that the gas path length is associated either with the working distance or with the distance between an on-axis pressure-limiting (differential) aperture and the specimen (refer back to Figure 4.1), then an apparently logical conclusion would be that it is better to maintain a short working distance in order to minimise primary electron scattering. However, there are a few factors to consider before placing a specimen very close to the lens or aperture, particularly if the latter forms part of a detector.

One observation is that there is a low pressure region just below the final aperture. Instabilities in the gas flow, where two different pressure zones meet, set a limit on how short the working distance should be. A general rule of thumb is that the working distance should be equal to or greater than the diameter of the final pressure-limiting (differential) aperture (Danilatos, 1994b). For example, if the aperture diameter is 500 μm (0.5 mm), then the specimen should be no closer than this. If the signal amplification mechanism is gas ionisation, then the acceleration and hence amplification of secondary electrons requires a certain amount of space, as described in Chapter 3, Section 3.3.2.2 (for steady-state swarm conditions) and so a small gap inhibits this important requirement.

A short working distance can be tolerated if the detector anode or light pipe is mounted a sufficient distance away, off-axis, as depicted in Figure 4.10. This ensures that, whilst minimising the primary beam gas path length, there is still an adequate cascade gas path length for any amplification processes. Note that the presence of a pressure-limiting (differential) aperture below the objective lens polepiece is optional. Without the aperture, the field of view is increased from a few hundred micrometres to a few millimetres, although the maximum pressure is then typically restricted to about 150 Pa (~1.1 torr).

The scenario in Figure 4.10 works well enough for low-voltage imaging. However, for X-ray microanalysis, the working distance (from the objective lens to the specimen) is a fixed length, say 10 mm, in order to maintain a sufficient solid angle for collection of X-rays as they take off towards the detector. Hence, it is not possible to substantially change

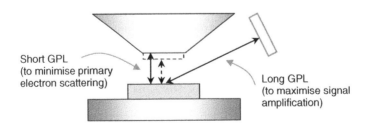

Figure 4.10 Schematic diagram showing a short working distance/short gas path length to minimise primary electron scattering, with a longer path length to allow adequate signal formation. The dashed line indicates that the pressure-limiting (differential) aperture is optional

the working distance and still obtain good quality X-ray spectra and elemental maps. However, this constraint can be considerably eased if we can decouple the working distance and gas path length. A method for this is shown in the next section.

Finally, the specimen or electron column may be at risk if the lens/aperture–specimen distance is very short. For example, consider a soft specimen such as a viscous emulsion, placed less than 0.5 mm beneath an on-axis pressure-limiting (differential) aperture. This means that the specimen is within the low-pressure turbulent region, where it may be pulled up into the column, causing contamination. Similarly, if the specimen contains water and has been suitably stabilised using the appropriate chamber conditions (Chapter 3, Section 3.5), then placing it in the low-pressure region may lead to unwanted evaporative loss of moisture.

4.4.5 Separating Gas Path Length from Working Distance

One simple solution to the issues outlined in the previous section is to artificially reduce the gas path length by fitting some form of tube to the end of the polepiece, extending down to just above the specimen surface. A built-in aperture at or near the end of the tube acts as the final pressure-limiting (differential) aperture and, in this way, the gas path length can be reduced to just 1 or 2 mm, while the specimen is a distance of 10 mm or more from the lens, as shown in Figure 4.11.

Using an arrangement such as that shown in Figure 4.11 means that primary beam electrons are quite well protected from the gaseous environment for several precious additional millimetres before encountering any significant quantity of gas molecules and the inherent increase in scattering.[3] This is a considerable improvement for X-ray analysis (see Section 4.6) and low-voltage imaging, and by cutting down on the number of electrons lost into the skirt, also boosts the signal-to-noise ratio for imaging (in other words, it helps to restore the useful primary beam current).

Figure 4.12 demonstrates three scenarios that arise from consideration of the working distance and gas path length. Note that the backscattered

---

[3] Inevitably, a small number of gas molecules will give rise to scattering even before the primary electrons enter the aperture or extension tube, but their effect is negligible under the circumstances discussed here.

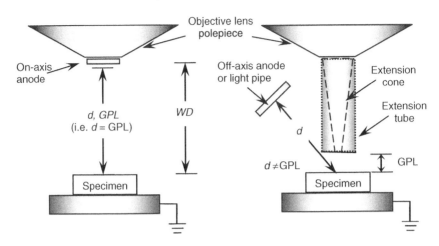

Figure 4.11 Schematic diagram showing the use of an extension tube (right-hand diagram) to overcome the limitations on working distance for the purposes of X-ray microanalysis, for example. The working distance can remain at a fixed value while, on the right, the gas path length GPL is significantly reduced. Note that there is a pressure-limiting (differential) aperture within the on-axis anode and at the end of the extension tube closest to the specimen

electron signal has been used to collect the results shown in Figure 4.12, to minimise the dependence of imaging on VP-ESEM signal amplification mechanisms (i.e. electron, ion and photon signals) and hence help us to get a better idea of how the primary electrons are affected by the changes in gas path length.

4.5 HOW MUCH GAS?

4.5.1 Introduction

The pressure range in the VP-ESEM can be varied from about 10 Pa up to 2660 Pa (20 torr).[4] With such a large range, it is important to choose the pressure appropriate to the experiment. In the simplest case, for imaging the surface of a nonvolatile electrically insulating material, the pressure should essentially be just sufficient to provide enough positive ions to counterbalance the negative charge implanted by the primary electron beam, whilst producing adequate signal carriers for imaging. However,

---

[4] This range is even larger in the very latest instruments, making it possible to go to pressures as high as 4 kPa (30 torr) for gases such as nitrogen.

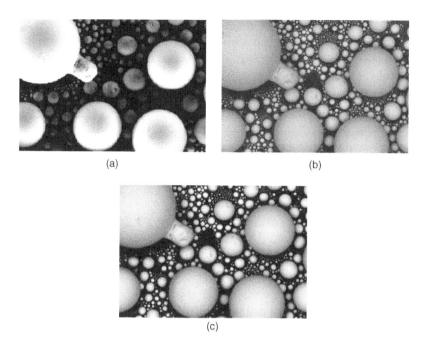

(a) (b)

(c)

Figure 4.12 Backscattered electron images to show the effect of (a) short working distance, short gas path length (3 mm), (b) long working distance, long gas path length (10.5 mm) and (c) long working distance (10.5 mm), short gas path length (3 mm). Notice how in (c) the contrast and signal-to-noise have improved. Imaged in nitrogen gas with primary beam energy $E_0 = 20$ keV. Horizontal field width = 255 μm. Images courtesy of Ken Robinson, Carl Zeiss SMT Ltd

as we will continue to see, this is not trivial, due to the interdependence of imaging parameters, not to mention the dielectric properties of the specimen itself (see also Chapter 5). The situation becomes even more complex when the specimen is moist or liquid, as this places additional constraints on the requisite pressure range (Chapter 3, Section 3.5).

The following sections aim to give an appreciation of the factors involved in choosing a gas pressure for a given specimen or experiment. However, a note of caution is needed here. For most experiments, the pressure need only be a few tens to a few hundreds of pascals (from a fraction of a torr up to one or two torr or so), depending on the dielectric properties of the specimen (i.e. a particularly nonconductive specimen will call for a greater concentration of ions).

For experiments involving transitions between the vapour and liquid states of water, the pressure range is generally higher, maybe 400–800 Pa (3–6 torr) depending on the temperature used. In some cases, the information presented in this chapter spans the pressure range all the way up to 2.66 kPa (~20 torr). But we only consider the effects that such pressures have on primary electrons as they travel to the specimen surface.

We do not consider the (generally negative) implications that high pressures have on the detection of signals. Suffice it to say that working at high pressures only makes sense if that is the regime required by the experiment, perhaps due to temperature–pressure considerations, and probably requires use of the measures described in Section 4.4.5, along with detectors specially suited to the purpose. An example will follow in Chapter 6.

4.5.2 Scattering of Primary Electrons as a Function of Pressure

4.5.2.1 Effect of Chamber Pressure on the Elastic Mean Free Path

We now consider the elastic scattering of primary electrons as a function of pressure over a range of primary electron beam energies E_0, for 5 keV $\leq E_0 \leq 30$ keV. In this section, we will use nitrogen as a model gas, to show the trends in behaviour. As before, we first consider the elastic mean free paths λ_e of primary electrons, calculated from Equations (2.6), (2.7) and (4.1), but this time for several primary beam energies over a range of gas pressures. As before, Recall that we have made a number of simplifying assumptions (Section 4.2), and so the numerical values shown should only be taken as an estimate.

Accordingly, Figure 4.13 is a log-log plot showing values of the elastic mean free path λ_e for nitrogen gas pressures up to 1 kPa (7.5 torr). We see that for pressure $p = 10$ Pa (0.075 torr), the mean free path for electrons having $E_0 = 10$ keV is around 2 cm, dropping by an order of magnitude for each order of magnitude increase in pressure. If the primary beam energy is increased to 30 keV, the values of λ_e increase to approximately 20 cm at $p = 10$ Pa, ~2 cm at $p = 100$ Pa and ~2 mm at $p = 1$ kPa.

Figure 4.14, meanwhile, has been plotted in order to see how the mean free path λ_e varies over the entire VP-ESEM pressure range, which is typically 2.66 kPa (20 torr). Again, the gas is nitrogen and a range of primary beam energies is shown. Notice that, even at the maximum pressure, the elastic mean free path for primary electron energy $E_0 = 30$ keV is $\lambda_e \approx 0.5$ mm. However, for $E_0 = 5$ keV, $\lambda_e \approx 20$ µm.

Ultimately, consideration of the mean free path in conjunction with a small gas path length (such as 1–2 mm) would suggest that, for nitrogen, a pressure $p = 100$ Pa or so should give a good-quality beam

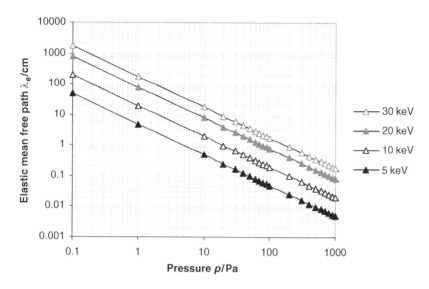

Figure 4.13 Log-log plot to show the variation in elastic mean free path λ_e for primary electrons in nitrogen gas as a function of pressure and a range of primary beam energies E_0

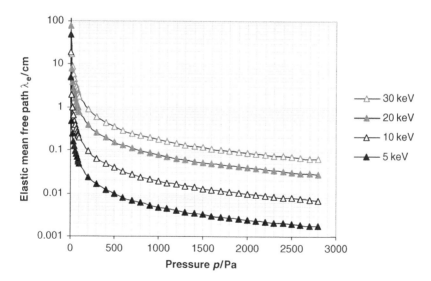

Figure 4.14 Log-linear plot to show the variation in elastic mean free path λ_e for primary electrons in nitrogen gas over the wide range of pressures available in the VP-ESEM and for several primary beam energies E_0

profile and that even several hundred pascals would maintain a primary electron scattering profile suitable for practical purposes, particularly for energies above about 10 keV. This is indeed found in practice. For lower energies, the degree of scattering will start to have a pronounced effect on the fraction of electrons remaining in the focused probe beyond $p \sim 100$–200 Pa. We will quantify these ideas in Section 4.5.2.3.

4.5.2.2 Influence of Pressure on the Radius of the Primary Beam Skirt

Let us look again at the size of beam skirt for a range of gases, this time plotting Equation (4.8) as a function of pressure and primary beam energy $E_0 = 10$ keV. We will use the same range of gases and a gas path length GPL = 2 mm. The resulting relationship between skirt radius r_s and pressure p is shown in Figure 4.15.

As we would expect, the skirt radius for helium increases slowly with increasing pressure, reaching $r_s \sim 5\,\mu$m at 500 Pa, while for argon this figure is nearer 40 μm.

For completeness, Figure 4.16 shows the general trend across an extended pressure range for the same boundary conditions as for Figure 4.14. At the maximum pressure shown (2.8 kPa), the skirt radius for argon is very large ($\sim$180 μm) and almost certainly corresponds to plural scattering and hence loss of the focused central probe.

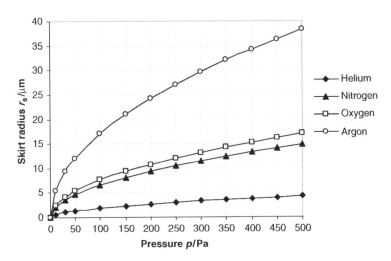

Figure 4.15 Plot to show skirt radii as a function of pressure for a range of gases having atomic numbers $Z = 2$ (helium), $Z = 7$ (nitrogen), $Z = 8$ (oxygen) and $Z = 18$ (argon). Primary beam energy $E_0 = 20$ keV and gas path length GPL = 2 mm

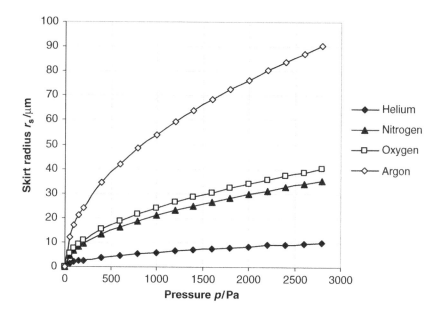

Figure 4.16 Plot of skirt radii r_s over the pressure range extending to 2.8 kPa for gases having atomic numbers $Z = 2$ (helium), $Z = 7$ (nitrogen), $Z = 8$ (oxygen) and $Z = 18$ (argon). Primary beam energy $E_0 = 20$ keV and gas path length GPL = 2 mm

Meanwhile, recall from Chapter 3, Section 3.2.2 that when the primary beam skirt is quite large (a few tens of microns, say), the delocalised primary electrons can spread to fill the whole field of view irrespective of the position of the focused probe. In this case, the skirt adds an essentially unvarying background component to the overall signal.

We know from Section 4.4 that reducing the gas path length can have a significant effect on the radius of the beam skirt, so we now look at the consequences of a range of gas pressures for different gas path lengths. Starting with nitrogen gas and a primary beam energy $E_0 = 20$ keV, Figure 4.17 shows the effect on skirt radius for several gas path lengths, 1 mm $\leq$ GPL ≤ 15 mm.

When the gas path length is kept at GPL = 1 mm, the skirt radius varies relatively little across the pressure range shown. Increasing to GPL = 5 mm immediately results in quite a significant increase: about a factor of 10. For example, for $p = 100$ Pa, $r_s = 26$ μm compared with $r_s = 2.4$ μm for GPL = 1 mm. Similarly, for $p = 2.8$ kPa, $r_s = 139$ μm for GPL = 5 mm, compared to $r_s = 12.5$ μm for GPL = 1 mm.

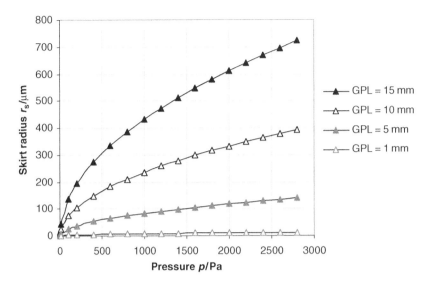

Figure 4.17 Plot of primary beam skirt radii r_s over the pressure range extending to 2.8 kPa for several gas path lengths and in nitrogen gas, $Z = 7$. Primary beam energy $E_0 = 20$ keV

If the gas path length GPL = 15 mm, these figures become ~140 and ~730 μm for $p = 100$ Pa and 2.8 kPa, respectively. In the latter case, it is highly unlikely that sufficient electrons remain in the focused probe.

In Figure 4.18, we explore a larger range of parameters with a log-log plot of skirt radius for the usual range of gases, over the extended pressure range to 2.8 kPa, for two different primary beam energies, E_0 = 1 keV and 30 keV. In addition to the overall trends that we might expect, we see that increasing the primary beam energy from 1 keV to 30 keV results in a decrease in beam skirt radius of roughly two orders of magnitude.

4.5.2.3 Influence of Pressure on the Useful Primary Electron Signal

The plot shown in Figure 4.19 gives the relationship between primary beam energy and electrons that remain in the focused probe, again in accordance with the arguments put forward in Section 4.2, for the now-familiar range of gases. The gas path length GPL = 2 mm, and primary beam energy $E_0 = 20$ keV.

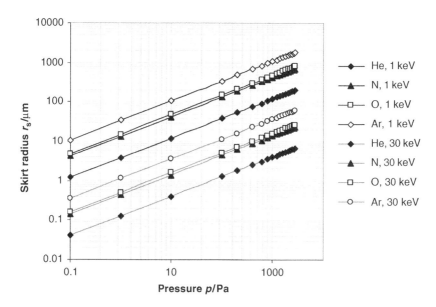

Figure 4.18 Log-log plot to show skirt radii over an extended pressure range for gases having atomic numbers $Z = 2$ (helium), $Z = 7$ (nitrogen), $Z = 8$ (oxygen) and $Z = 18$ (argon) for two primary beam energies $E_0 = 1$ and $30\,keV$. Gas path length GPL = 2 mm

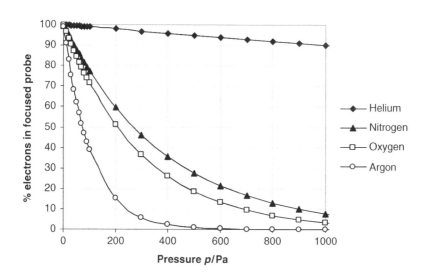

Figure 4.19 Plot to show the percentage of electrons remaining in the focused probe, indicating the useful primary current, as a function of pressure and a range of primary beam energies. Gas path length GPL = 2 mm, primary beam energy E_0 = 20 keV

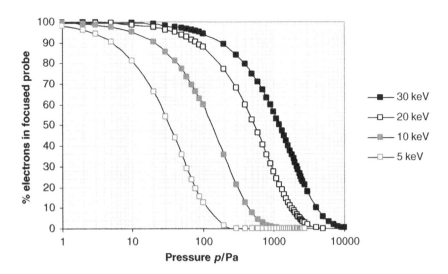

Figure 4.20   Log-linear plot to show the percentage of electrons remaining in the focused probe as a function of pressure and a range of primary beam energies for nitrogen gas, $Z = 7$. Gas path length GPL = 1 mm

Again, we see that there is very little scattering of helium, due to its very low atomic number and correspondingly low atomic weight and molar mass. For nitrogen and oxygen, the results suggest that pressures of a few hundred pascals will still give an adequately useful primary current, particularly below about 200 Pa, while for argon, the pressure range is rather more restrictive under these conditions.

Finally, Figure 4.20 is a log-linear plot to show what happens to the useful primary electron beam current for gas path length GPL = 1 mm and nitrogen gas, $Z = 7$, across the extended pressure range for a range of primary beam energies $5 \leq E_0 \leq 30$ keV. According to these data, primary beam energy has a very strong influence on useful current.

As predicted earlier (Section 4.5.2.1), the fraction of focused electrons remains relatively high for beam energies above 10 keV and pressures of a few hundred pascals. For energies around 5 keV, formation of a focused probe is highly unlikely at the highest pressure, but still possible for high energies.

So those were the basics of electron scattering in VP-ESEM. Our next concern is the effect that these parameters have on chemical analysis in a gaseous environment, as we shall see in Section 4.6.

4.6 X-RAY MICROANALYSIS IN VP-ESEM

4.6.1 Introduction

If properly controlled, the diffuse skirt of scattered primary electrons can have very little noticeable effect on images in VP-ESEM. With a suitable choice of gas type, pressure and gas path length, image resolution is not generally compromised. However, a certain amount of primary electron beam current may be lost, thus the signal-to-noise ratio is decreased and a uniform background signal is added.

But the effects of the gas and of the skirt electrons are felt much more strongly when trying to perform X-ray microanalysis. Here, the spatial resolution can be considerably compromised, since X-rays may be generated tens or hundreds of microns away from the impact point of the focused beam. This leads to inaccuracies between the intended point of reference (the beam impact point) and the real position of a given element when analysing X-ray data.

It should be said, though, that the possibility to image and chemically analyse specimens in the absence of both specimen coatings and negative charging is a great benefit, certainly for rapid, qualitative work, subject to the proviso that the surface has not developed a positive potential due to the presence of excess positive ions and that an adequate gas pressure has been employed to avoid any negative potential (Chapter 5).

However, it is, of course, important to be aware of the limitations imposed by the presence of the gas, and to take steps to minimise the effects. The following sections summarise the main points to bear in mind. An excellent review of this aspect of VP-ESEM is given by Newbury (2002), and there are numerous other articles in the literature, of which a small selection is given here: Egertonwarburton et al. (1993); Danilatos (1994a); Gilpin and Sigee (1995); Mansfield (2000); Griffin and Suvorova (2003); Carlton et al. (2004); Khouchaf and Boinski (2007); Le Berre et al. (2007a).

4.6.2 Effects of Chamber Gas on X-ray Signals

When carrying out X-ray microanalysis, there is evidence for both elastic and inelastic scattering of the primary beam. The formation of the beam skirt is assumed to be largely due to elastic scattering, and this was used as the basis for the modelling of primary electron mean free paths, skirt radii and useful electron currents in previous sections.

Inelastic interactions, meanwhile, can be detected by virtue of X-ray signals, one of the many beam–specimen products discussed in Chapter 2. Spurious (i.e. unwanted) X-ray signals can arise from collisions of primary beam electrons with chamber gas molecules, and also from inelastic interactions of emitted backscattered electrons as they travel through the gas away from the specimen.

As an example of the relative significance of this effect, Newbury (2002) showed that for a carbon disc irradiated by primary electrons with $E_0 = 20\,keV$, gas path length GPL = 6 mm and with water vapour as the chamber gas, there is no appreciable effect of the gas at pressures below 50 Pa (0.4 torr). This is in accord with the findings in Section 4.5, where electron mean free paths exceed the gas path length at very low pressures, and so we expect that there will be very little primary electron scattering.

However, as Newbury discusses, two effects start to affect the results at water vapour pressures above about $p = 133$ Pa (1 torr). The first is that oxygen counts are recorded and a corresponding peak appears in the X-ray spectrum, even though this element is not present in the specimen. These are spurious signals generated by inelastic collisions of electrons with water molecules in the gas. The second effect is that the target X-ray count (hence peak intensity) is reduced due to scattering of primary electrons into the increasingly tenuous skirt, producing X-rays outside the acceptance angle of the X-ray detector. An example of how this effect can manifest itself is shown in Figure 4.21.

A similar experiment to that outlined above was carried out using a hydrogen–helium mixture in place of water vapour. These elements give no measurable intrinsic characteristic X-ray emission and, as we have seen, light elements produce the least primary electron scattering. Hence, under the same conditions as described for water vapour above, extraneous X-rays from the gas are eliminated. Note that this gas mixture does still contribute to the overall spectrum due to continuum (background) X-ray emission.

Meanwhile, according to theoretical predictions, absorption of X-rays by the gas is negligible at the pressures typically used for nonhydrated specimens in VP-ESEM, e.g. up to a couple of hundred pascals (i.e. up to 1.5 torr or so), but would be very significant at the maximum chamber pressure (2.66 kPa, 20 torr) (Newbury, 2002).

Next, if we extend these ideas to the microanalysis of a heterogeneous specimen, we find that X-rays may be generated by the skirt electrons

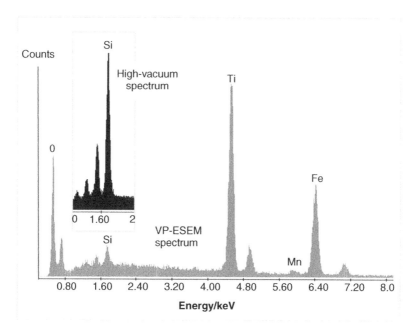

Figure 4.21 Spectra showing the general effect of acquiring X-ray data in a gaseous environment vs high vacuum. Qualitatively, the information is similar in both cases, but for the VP-ESEM spectrum, the peak intensity is reduced. Courtesy of Dirk van der Wal, FEI Company

at features perhaps far removed from the impact point of the beam (see, for example, Carlton, 1997). If the elemental compositions differ between these regions, X-ray data may be collected for elements not actually present at the impact point, which will then simultaneously and erroneously appear in the spectrum recorded for that point. The relationship between primary beam X-ray interaction volume and X-ray generation due to primary beam skirt electrons is depicted schematically in Figure 4.22.

4.6.3 Considerations for Minimising the Effects of the Gas

X-ray microanalysis places some constraints on the choice of parameters in the VP-ESEM. We have already dealt with the issue of the optimal working distance for collecting the X-ray signal (Section 4.4.5) and how to maintain this working distance while minimising the gas path length (Figure 4.11). Another consideration is the primary beam energy E_0. As with secondary electron emission, the efficiency of X-ray generation

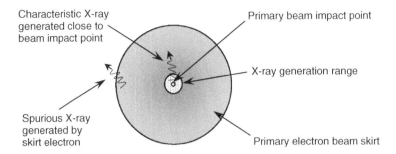

Characteristic X-ray generated close to beam impact point

Primary beam impact point

X-ray generation range

Spurious X-ray generated by skirt electron

Primary electron beam skirt

Figure 4.22 Interactions between the primary beam and specimen result in X-ray photon generation corresponding to the size of the interaction volume. However, as this generalised schematic diagram shows, primary electrons scattered into the skirt can lead to anomalies in the analysis of X-ray data, as X-rays can be emitted some tens or hundreds of microns from the impact point of the primary beam

is a function of beam energy. For X-ray emission, a useful rule of thumb is to use a beam energy about twice the critical energy E_c of the X-ray excitation of interest, in order to maintain high efficiency and also generate sufficient characteristic X-rays above the continuum background. Now, a primary beam energy $E_0 = 20\,\text{keV}$ gives good excitation of the upper part of the X-ray photon energy range for most elements and, as we know, helps to reduce primary electron scattering in the gas. But this is at the expense of the low energy X-ray intensity which, because these X-rays are generated at large depths for high primary beam energies, is absorbed within the material. Thus, if the X-ray edge in question happens to be at an energy $E_c \approx 2\,\text{keV}$, ideally one would work at $E_0 = 4\,\text{keV}$ to maximise the intensity, but this introduces a vastly increased risk of primary electron scattering. Clearly, when using a low primary beam energy, minimising the gas path length and gas pressure (whilst compensating for charging) is a very good practice to adopt.

Whilst choosing a gas pressure, care should be taken not to allow electrical potentials to develop since, if the specimen acquires a net negative or positive potential as will be outlined in Chapter 5, Section 5.4, the landing energies of primary electrons can be altered, causing a shift in the Duane–Hunt limit. It will be shown in Section 5.4.2 that this can be used as a method for measuring the sign and magnitude of surface potential. In the case of negative charging, the primary beam landing energy is lower than the energy defined by the user and so X-ray excitation diminishes towards the high-energy end of the X-ray energy spectrum. Counts are reduced and peaks start to disappear.

As we know from Sections 4.4 and 4.5, helium gas causes very little scattering of the primary beam, and so it could be thought of as a very useful gas for microanalysis. However, helium is not without its drawbacks (mentioned briefly in Section 4.3.4.3), the chief of which is the difficulty of pumping such small atoms in the vacuum system. An additional restriction may be that the choice of chamber gas is determined by the experimental requirements. For example, water vapour is essential when dealing with moist or liquid specimens, and the gas pressure will necessarily be a few hundred pascals (or roughly 3–6 torr), depending on specimen temperature. So let us assess a simple alternative strategy. We have seen that the gas path length can be varied and that a short gas path length gives significant benefits.

Figure 4.23 shows a direct comparison between the skirt radii for helium and nitrogen for two values of gas path length, GPL = 10 mm and 1 mm, with primary beam energy $E_0 = 10$ keV. For GPL = 10 mm, there is a large disparity between nitrogen and helium. However, if the gas path length is reduced to 1 mm, the skirt radii become small and less sensitive to the effects of increasing pressure. Helium may still be the better choice for high resolution work, but other gases such as nitrogen, air and water vapour make excellent alternatives for most routine work, provided that the gas path length is suitably minimised.

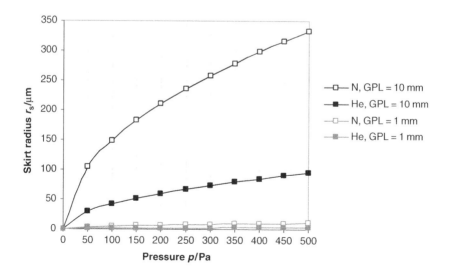

Figure 4.23 Plot of skirt radii for nitrogen and helium for two different gas path lengths, GPL = 1 mm and 10 mm. Primary beam energy $E_0 = 10$ keV. When the path length is minimised, primary beam scattering in the two gases becomes comparable

4.6.4 Post-Acquisition Methods to Correct for Scattering

We know that we have to live with a certain amount of scattering in VP-ESEM. Its effects mean that the matrix corrections usually applied in quantitative microanalysis become inaccurate. However, several methods have been proposed to correct for the contributions of the skirt electrons after the data have been collected. A pressure variation method (Doehne, 1997) involves collecting spectra for two pressures, with all other parameters constant, and uses a difference spectrum to extrapolate back to the zero-scattering case. Consider a spectrum, A, recorded at a pressure p_A and another, spectrum B, recorded at a lower pressure p_B, the spectrum C that might be obtained in the absence of scattering is estimated from Equation (4.10):

$$C = B - [(A - B) \cdot d] \qquad (4.10)$$

where d is an empirical scaling factor given by $d = p_B/p_A$.

The method assumes that the compositional information from skirt-induced X-rays does not change with pressure: it is only the intensity that changes. This means that as the extent of features enveloped by the skirt changes, the relative X-ray contribution from each constituent is expected to remain in the same proportion. This may or may not be a significant factor, depending on the distribution and sizes of different phases in the specimen. If the gas path length is short (say, 1 mm), then we have already seen that the change in skirt radius varies slowly with pressure, in which case the assumption holds well. A similar method was described by Bilde-Sorensen and Appel (1997).

More recently, another pressure variation method has been proposed (Gauvin, 1999), which extrapolates to the zero-scattering X-ray intensity I as a function of the unscattered and scattered electron beam intensities, I_p and I_m, respectively, via Equation (4.11):

$$I = (I_p - I_m) \cdot f_p + I_m \qquad (4.11)$$

This can be further extended to deal with measurements at two different pressures (Le Berre et al., 2007b).

Other methods include utilising the continuum (Bremsstrahlung) spectrum to restore peak intensity (Griffin and Nockolds, 1996), and a beam stop method for separating the contributions generated by the focused electron beam from those of the skirt that is particularly useful for work involving higher pressures, such as those needed for hydrated specimens (Bilde-Sorensen and Appel, 1997).

Another interesting idea is to use X-ray focusing optics. A tapered polycapillary optic can propagate X-rays along its length via multiple low-angle reflections, and the tapering means that the X-rays can be made to follow a converging or diverging path. The capillary is very sensitive to photon energy and angle and could therefore be used to exclude X-rays produced remotely by the skirt and only collect those generated by the focused beam (see Newbury, 2002 and references therein).

REFERENCES

Bilde-Sorensen, J. and Appel, C.C. (1997). *X-ray spectrometry in ESEM and LVSEM: corrections for beam skirt effects*. SCANDEM-97.

Carlton, R.A. (1997). The effects of some instrument operating conditions on the x-ray microanalysis of particles in the environmental scanning electron microscope. *Scanning*, **19**, 85–91.

Carlton, R.A., Lyman, C.E. and Roberts, J.E. (2004). Charge neutralization in the ESEM for quantitative X-ray microanalysis. *Microsc. Microanal.*, **10**(6), 753–763.

Danilatos, G.D. (1988). Foundations of Environmental Scanning Electron Microscopy. *Adv. Electron. Electr. Phys.*, **71**, 109–249.

Danilatos, G.D. (1994a). Environmental Scanning Electron-Microscopy And Microanalysis. *Mikrochim. Acta*, **114**, 143–155.

Danilatos, G.D. (1994b). Environmental scanning electron microscope: Some critical issues. *Scanning Microscopy, Supplement 7, 1993 – Physics Of Generation And Detection Of Signals Used For Microcharacterization*, 57–80.

Doehne, E. (1997). A New Correction Method for High-Resolution Energy-Dispersive X-Ray Analyses in the Environmental Scanning Electron Microscope. *Scanning*, **19**(2), 75–78.

Egertonwarburton, L.M., Griffin, B.J. and Kuo, J. (1993). Microanalytical Studies Of Metal Localization In Biological Tissues By Environmental SEM. *Microsc. Res. Techn.*, **25**(5–6), 406–411.

Fletcher, A., Thiel, B. and Donald, A. (1997). Amplification measurements of Potential Imaging Gases in Environmental SEM. *J. Phys. D: Appl. Phys.*, **30**, 2249–2257.

Gauvin, R. (1999). Some theoretical considerations on x-ray microanalysis in the environmental or variable pressure scanning electron microscope. *Scanning*, **21**(6), 388–393.

Gillen, G., Wight, S., Bright, D. and Herne, T. (1998). Quantitative secondary ion mass spectrometry imaging of self-assembled monolayer films for electron beam dose mapping in the environmental scanning electron microscope. *Scanning*, **20**, 400–409.

Gilpin, C. and Sigee, D.C. (1995). X-Ray-Microanalysis Of Wet Biological Specimens In The Environmental Scanning Electron-Microscope.1. Reduction Of Specimen Distance Under Different Atmospheric Conditions. *J. Microsc.–Oxford*, **179**, 22–28.

Goldstein, J., Newbury, D., Joy, D., Lyman, C., Echlin, P., Lifshin, E., Sawyer, L. and Michael, J. (2003). *Scanning Electron Microscopy and X-Ray Microanalysis*, third edition. Plenum.

Griffin, B.J. and Nockolds, C.E. (1996). *Quantitative EDS analysis in the ESEM using a bremsstrahlung intensity-based correction for primary electron beam variation and scatter*. Microscopy and Microanalysis '96.

Griffin, B.J. and Suvorova, A.A. (2003). Charge-related problems associated with X-ray microanalysis in the variable pressure scanning electron microscope at low pressures. *Microsc. Microanal.*, 9(2), 155–165.

Kadoun, A., Belkorissat, R., Khelifa, B. and Mathieu, C. (2003). Comparative study of electron beam–gas interaction in an SEM operating at pressures up to 300 Pa. *Vacuum*, 69(4), 537–543.

Khouchaf, L. and Boinski, F. (2007). Environmental Scanning Electron Microscope study of SiO_2 heterogeneous material with helium and water vapor. *Vacuum*, 81(5), 599–603.

Le Berre, J.F., Demers, H., Demopoulos, G.P. and Gauvin, R. (2007a). Examples of charging effects on the spectral quality of X-ray microanalysis on a glass sample using the variable pressure scanning electron microscope. *Scanning*, 29(6), 270–279.

Le Berre, J.F., Demopoulos, G.P. and Gauvin, R. (2007b). Skirting: A limitation for the performance of X-ray microanalysis in the variable pressure or environmental scanning electron microscope. *Scanning*, 29(3), 114–122.

Mansfield, J.F. (2000). X-ray microanalysis in the environmental SEM: A challenge or a contradiction? *Mikrochim. Acta*, 132(2–4), 137–143.

Mathieu, C. (1999). The beam–gas and signal–gas interactions in the variable pressure scanning electron microscope. *Scanning Microsc.*, 13(1), 23–41.

Morgan, S.W. and Phillips, M.R. (2006). Gaseous scintillation detection and amplification in variable pressure scanning electron microscopy. *J. Appl. Phys.*, 100(7), Article no. 074910.

Newbury, D.E. (2002). X-ray microanalysis in the variable pressure (environmental) scanning electron microscope. *J. Nat. Inst. Standards Technol.*, 107(6), 567–603.

NIST Atomic Spectra Database, http://physics.nist.gov/PhysRefData/ASD/index.html.

Reimer, L. (1985). *Scanning Electron Microscopy. Physics of Image Formation and Microanalysis*. Springer-Verlag.

Stowe, S.J. and Robinson, V.N.E. (1998). The use of helium gas to reduce beam scattering in high vapour pressure scanning electron microscopy applications. *Scanning*, 20, 57–60.

Tang, X.H. and Joy, D.C. (2005). An experimental model of beam broadening in the variable pressure scanning electron microscope. *Scanning*, 27(6), 293–297.

Thiel, B.L., Bache, I.C. and Smith, P. (2000). Imaging the Probe Skirt In the Environmental SEM. *Microsc. Microanal.*, 6(Suppl. 2), 794–795.

Thiel, B.L., Toth, M., Schroemges, R.P.M., Scholtz, J.J., van Veen, G. and Knowles, W.R. (2006). Two-stage gas amplifier for ultrahigh resolution low vacuum scanning electron microscopy. *Rev. Sci. Instr.*, 77(3).

Wight, S., Gillen, G. and Herne, T. (1997). Development of environmental scanning electron microscopy electron beam profile imaging with self-assembled monolayers and secondary ion mass spectroscopy. *Scanning*, **19**, 71–74.

Wight, S.A. and Zeissler, C.J. (2000). Direct measurement of electron beam scattering in the environmental scanning electron microscope using phosphor imaging plates, *Scanning*, **22**(3), 167–172.

5

Imaging Uncoated Specimens in the VP-ESEM

5.1 INTRODUCTION

So far, we have necessarily concentrated on operational aspects of the SEM and VP-ESEM. Now we turn our attention to the implications of working with uncoated specimens by considering some of the physical properties of materials that are relevant parameters in the operation of the VP-ESEM (also applicable to the SEM) and the influence of the specimen itself as an integral part of the system. It is particularly important to have a grasp of the ways in which electrons move around in and escape from materials, and how certain materials become electrically charged, both in high-vacuum conditions and in the gaseous environment of the VP-ESEM.

Since electronic structure plays such a vital role in determining the diversity of the properties of matter, some of the basic electronic properties of conducting, semiconducting and nonconducting materials will be outlined, so that we can appreciate how charge carriers behave in and around these materials. It is also useful to know a little about the various electric fields that arise in and around the specimen, in order to work towards an understanding of how to maintain control over a specimen with given electronic properties and so determine the most appropriate set of conditions for imaging and microanalysis. Crucially, this means that we must assess the role of positive ions in the VP-ESEM, which can have both beneficial and detrimental effects on imaging.

Principles and Practice of Variable Pressure/Environmental Scanning Electron Microscopy (VP-ESEM)
D. J. Stokes
© 2008 John Wiley & Sons, Ltd

Other factors to consider when imaging uncoated specimens, especially for soft matter and hydrated materials, include the penetration of the primary beam as a function of energy and the risk of radiation damage, particularly in a water vapour environment.

5.2 ELECTRONIC STRUCTURE

5.2.1 The Energy Level Diagram

In the case of a single atom, electrons occupy orbitals which have discrete energy levels, or states. In other words, electrons can only have certain distinct energies, and these energies are atom-specific. Meanwhile, when many atoms (or molecules) are brought together, the atomic (or molecular) orbitals get close enough to form energy bands. The widths of these bands, and any overlap or separation, are key to determining the electronic properties of the material or substance.

Some electrons occupy energy states that are more tightly bound than others. Those in the outer shells are termed valence electrons and, in certain materials, one or two valence electrons per atom are able to move freely within the material, forming a 'sea' around their positive ion cores. Note that a detailed discussion of the wave functions associated with electron orbitals and ionic cores is beyond the scope of this book, and the reader is referred to, for example, Kittel (1986).

Those valence electrons that can move around freely are in fact called conduction electrons, and occupy the conduction band E_c (see Figure 5.1). The remaining valence band electrons occupy the valence band E_v. At the highest level, the electronic states are so close together

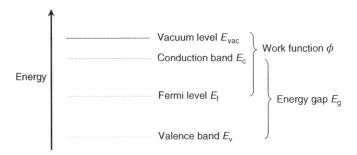

Figure 5.1 Generalised schematic energy level diagram showing the various electronic bands and levels associated with the overlapping of atomic orbitals. See text for discussion

that there is effectively a continuum of states in which electrons are no longer bound, and hence they can escape the material. This is termed the vacuum level E_{vac}.

Importantly, forbidden energy gaps exist in the band structures of insulators and semiconductors, which are absent in the case of metallic materials. We will come back to the reasons for this shortly but, suffice it to say, this property has a profound effect on the mobility of electrons in these materials when an electric field is applied. In addition, various trapping states may exist or arise that can alter the characteristics of the electronic structure.

Now, given that a material contains only a certain number of electrons and many energy states, the probability of particular electron states being filled at a given temperature is given by Fermi–Dirac statistics (the details need not concern us here). The natural energy at which this occurs is taken to be within the energy gap for insulators and semiconductors, and is called the Fermi level E_f. For metals, the position of the Fermi level is just above the highest occupied level in the valence band. Finally, the potential difference between the vacuum level and the Fermi level is called the work function Φ, and is effectively a barrier to reaching the electron energy required to escape the material. A convenient way to visualise electronic structure is via an energy level diagram, as shown in Figure 5.1.

5.2.2 Conductors, Semiconductors and Insulators

An electrically conductive (i.e. metallic) material is one for which the conduction band is empty and the valence band partially filled, and these may overlap, as shown in Figure 5.2(a), or be separated by a small gap. In either case, electrons are free to move around when an electric field is applied, and hence metals conduct electricity easily. If these electrons are given sufficient energy to overcome the potential barrier at the surface (the work function Φ) then their energy takes them above the vacuum level E_{vac} and they can leave the material, as mentioned above.

Meanwhile the band structures of semiconducting materials differ from those of conductors in that the conduction and valence bands are partially filled. The bands are separated by an energy gap E_g and, for intrinsic semiconductors, a little additional thermal energy means that electrons can move between the bands. Once in the conduction band, electrons are free to move around due to the partially filled nature of the bands, in common with metals. The band structure of an intrinsic semiconductor is represented in Figure 5.2(b).

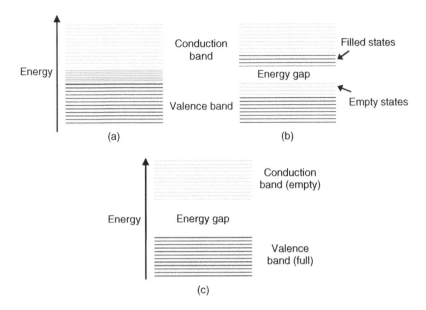

Figure 5.2 Schematic diagram showing the electronic band structure of (a) a metallic material, (b) an intrinsic semiconductor and (c) an insulator. In the metallic state, conduction and valence bands overlap, and charge can move around easily in an applied electric field. In (b), the conduction and valence bands are separated by a small energy gap, but empty states in each band make it relatively easy for electrons to move between the bands when energy is supplied. For a typical insulator, shown in (c), the bands are farther apart. There are no empty states in the valence band for electrons to move between; hence it is more difficult for conductivity to occur in insulating materials

Alternatively, for extrinsic semiconductors, dopants can be introduced into the material, resulting in the formation of discrete, localised impurity levels within the energy gap, as shown in Figure 5.3. Depending on the nature of the host material and dopant, impurity atoms act either to donate electrons to the conduction band or accept electrons from the valence band. These are termed n-type and p-type semiconductors, respectively.

Finally, the band structure of a typical insulator is shown in Figure 5.2(c). In this case the conduction band is empty and the valence band is full. This makes it very difficult for electrons to move in an applied electric field since, in order for conductivity to be a possibility, there must be energy states available within the same band. The energy gap for insulators is larger than for semiconductors (several eV), so that even with thermal excitation, electrons do not tend to traverse between the

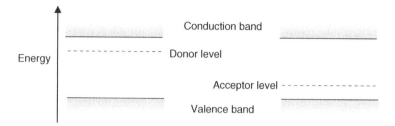

Figure 5.3 Schematic diagram to represent discrete, localised impurity levels for donor (*n*-type) and acceptor (*p*-type) impurities

bands. However, once promoted to the conduction band (perhaps due to a defect or impurity), electrons can move around and, again, escape if they have enough energy to overcome the surface potential barrier.

Factors that affect the size of the energy gap E_g include not only impurities such as the dopants discussed above but, more generally, the bonding between atoms in an element or compound. The number of bonds each atom makes with a neighbouring atom (known as bond order) has an effect on the associated electron orbital cloud, and this ultimately affects the size of the energy gap. For example, the presence of delocalised orbitals in double bonds (π-bonds, bond order $= 2$) results in a smaller energy gap compared to a material made up of only single bonds (σ-bonds, bond order $= 1$).

All of the various properties discussed have an effect on the way that primary, backscattered and secondary electrons behave within materials in the electron microscope, particularly at lower energies, determining the generation, transport and escape of electrons, and enabling us to understand the likelihood of electrons becoming trapped in a material and, if so, under what circumstances and for how long. This will be a recurring theme throughout this chapter, beginning in the next section, where we look at the behaviour of secondary electrons.

5.3 FACTORS AFFECTING SECONDARY ELECTRON EMISSION

5.3.1 Transport of Excited Electrons

The generation of secondary electrons was discussed in Chapter 2, Sections 2.4.1.2 and 2.4.3 and so will not be discussed further here, except to note that, strictly speaking, these should only be termed 'secondary' electrons after emission from the specimen. Until then, they are 'excited' electrons that can meet a number of alternative fates as they traverse the bulk material, depending on the prevailing energy states.

In metals, there is a continuum of energy states that can absorb the energy of excited electrons. These can range from ionisation events to collective oscillations of valence or conduction electrons (plasmons), all the way down to the production of heat (phonons). This explains why the mean free paths of excited electrons are so short in metallic materials and hence why the escape depths of secondary electrons are very small.

Now, as we saw in the previous section, other materials have some degree of discontinuity in their energy levels – the energy gap E_g – and this gap signifies a certain absence of mechanisms to absorb the energy of excited electrons. Put another way, excited electrons having energies within the range of the energy gap can travel greater distances without interacting with the material. Indeed, it is estimated that excited electrons require double, or perhaps 1.5 times, the energy of the gap in order to participate in anything other than phonon (heat) production (Bishop, 1974; Howie, 1995).

In turn, the electrical properties of semiconductors and insulators increase the mean free paths and escape depths of secondary electrons, and the extent to which this happens varies from material to material, depending on the size and energy range of the energy gap. To a first approximation the higher the bond order, the more delocalised the electron cloud, the smaller the energy gap.

In the case of insulating materials, the secondary electron mean free path characteristics require more sophisticated models than those for metals, and an example is that of Akkerman et al. (1996), which takes into account dielectric properties, energy gap and valence band width. Discussions relating this to specimens in the VP-ESEM can be found in Stokes et al. (1998) and Thiel and Toth (2005).

Additional states in the energy gap, such as impurities and local defects, effectively serve to further reduce the energy gap and increase the energy-absorbing events that excited electrons experience. These are important factors when considering secondary electron emission from uncoated materials in the SEM and VP-ESEM, and have been shown to yield information between regions with very subtle differences in electronic properties (a few examples include Chi and Gatos, 1979; Castell et al., 1997; Doehne, 1998; Stokes et al., 1998; Griffin, 2000; Elliott et al., 2002).

5.3.2 Escape of Excited Electrons

In order to escape from the sample surface, excited electrons must have sufficient energy to overcome the surface potential. For many solids, the escaping electron must overcome the work function Φ, as mentioned in Section 5.2.1, and typical values for metals are in the range 4–6 eV (Lide, 1991). Surface energy barriers in other materials (e.g. molecular substances) may be approximated by the electron affinity E_a (Howie, 1995), and their values tend to be lower than for metals. For example, the electron affinity for water is generally taken to be $E_a = 1.2$ eV.

But, for an excited electron in the VP-ESEM, this is far from the end of the story, for the electrons are not being released into vacuum as they would be in conventional high-vacuum SEM but, rather, into the more unusual environment of neutral and charged gas atoms or molecules. Now they must face a new set of challenges, not least of which is to avoid the effects of the otherwise useful positive ions waiting nearby. We will come back to this later, in Section 5.6.

5.4 THE INFLUENCE OF THE SPECIMEN ON THE SYSTEM

5.4.1 The Effect of Charging – the General Case

The depth to which primary electrons are implanted into a material decreases with atomic number Z and increases with primary beam energy E_0. The maximum implantation depth is known as the maximum electron range R_{max}, and so defines the limit at which negative charge due to primary electrons will be found within the specimen. At this limit, primary electrons are said to be 'thermalised'. An expression, due to Kanaya and Okayama (1972), gives the electron range R_{KO} as:

$$R_{KO} = \frac{0.0276A}{Z^{0.89}\rho} E_0^{1.67} \qquad (5.1)$$

where R_{KO} is measured in µm when using the constant 0.0276, $A =$ atomic weight (g/mole), $Z =$ atomic number, $\rho =$ density (g/cm^3), $E_0 =$ beam energy (keV).

Backscattered electrons are emitted from a relatively large region of the interaction volume (see Chapter 2, Figure 2.16), with a range of about one third of the maximum value of R_{max}, which gives some primary electrons a direct means of escaping the specimen. In a nonconductive material, charge that does not escape in this way can become trapped in the bulk for a finite period of time.

Meanwhile, close to the surface, excited electrons are emitted as secondary electrons from a small depth, determined by their mean free paths λ. The emission depth falls off sharply within the specimen, with a maximum range of roughly 5λ (Chapter 2, Section 2.4.3). The surface is thus depleted of electrons, leaving a net positive layer and, due to the electronic structure characteristics already discussed, the depth of this layer is much smaller in metals ($\sim1\,\text{nm}$) compared with insulators ($\sim20\,\text{nm}$) (Seiler, 1983).

This scenario is quite typical for high-vacuum SEM and, for poorly conducting materials, results in a dipole field inside the specimen. Recalling the general electronic structure for insulators outlined in Section 5.2, we note that electrons in the conduction band will be propelled towards the surface, while holes in the valence band will be drawn further into the bulk, in the direction of the dipole field. We can represent these concepts diagrammatically, as shown in Figure 5.4.

The net dipole field strength increases with increasing beam current and/or accumulated dose, decreases with increasing electron emission and is a function of the dielectric properties of the individual specimen or region.

One effect of the internal field due to negative charging is to exert a repulsion force on the incoming primary electrons. This deceleration causes primary electrons to land, not with their initial energy E_0, but with some other energy E_L (which will be lower than E_0 in this case).

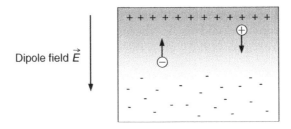

Figure 5.4 Schematic diagram showing the distribution of charge within a poorly conducting material in high-vacuum SEM. The separation of the charges leads to a dipole field between the surface and the bulk. If a negative charge is placed in this field, it will be attracted towards the surface. Similarly, a positive charge, or hole, will be drawn towards the bulk

A general expression for the landing energy E_L is given by Equation (5.2):

$$E_L = E_0 + eV_s \qquad (5.2)$$

where E_0 = initial primary beam energy in electronvolts eV, V_s = net surface potential in volts V (either positive or negative) and e is the charge of an electron.

Incidentally, the electronvolt is a rather confusing unit, and so a brief explanation of the units in Equation (5.2) is warranted. The energy E represented by the electronvolt is eV $\approx 1.6 \times 10^{-19}$ J, and the charge on the electron is $e \approx 1.6 \times 10^{-19}$ C. Meanwhile the volt has units V = J/C, and so can equivalently be expressed as V = eV/e (i.e. 1 V = 1.6 $\times 10^{-19}$ J/1.6 $\times 10^{-19}$ C = 1 J/C). Likewise, E = $e \cdot$ V = e(J/C) = eV.

We saw in Chapter 2 that secondary electron emission is more efficient with decreasing beam energy (see Section 2.4.3). Hence, if the primary electron landing energy E_L is reduced due to negative charging in the bulk and hence a net negative surface potential, the secondary electron yield increases and the signal becomes more intense. The effect of this increased intensity on the appearance of an image is a characteristic sign of negative charging. The worst case scenario is when the repulsion force is so high that primary electrons are completely deflected and strike the chamber walls, polepiece detectors and so on. An example of this was shown in Chapter 1 (the mirror effect, Figure 1.2). Conversely, if the specimen has a positive surface potential, brought about by having a landing energy somewhere between the first and second crossover energies E_1 and E_2 (see the discussion on low-voltage imaging in Chapter 2, Section 2.5.2), the electron emission intensity decreases and charged regions appear darker in the image.

During the imaging of poorly conductive materials, charge can accumulate in a single scan according to Equation (5.3) (after Shaffner and Van Veld, 1971):

$$\sigma_0 = I_0(1 - \eta - \delta)F/A \qquad (5.3)$$

where σ_0 = charge accumulated per unit area (C/m^{-2}), I_0 = primary beam current (A), η and δ are the backscattered and secondary electron coefficients, respectively, F = frame rate (s) and A = area scanned (m^2).

For a given amount of charge per unit area initially implanted σ_0, the charge σ_t remaining after a time t is given by Equation (5.4):

$$\sigma_t = \sigma_0 e^{-t/\tau} \quad \text{with } \tau = RC \qquad (5.4)$$

where τ is the time constant for charge decay, which is a function of the resistance R and capacitance C of the specimen (see also Thiel and Toth, 2005).

In effect, if the frame period F is smaller than the time constant τ, a poorly conducting material may accumulate charge over a number of successive frames, and for a specimen containing regions with differing values of τ (i.e. RC), some features will exhibit the effects of charging earlier than others.

Figure 5.5 shows schematically how, under high-vacuum conditions in the absence of any significant mechanisms for charge removal, charge gradually leaks away (decays), as shown in (a), but some charge remains as the primary electron beam returns to scan the next frame. The interval between frames is depicted in (b), while the overall effect is to

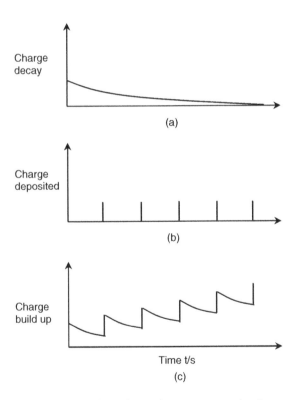

Figure 5.5   If charge is not conducted away between successive frames, charge builds up with each successive frame, ultimately precluding imaging with the electron beam. This is a less likely scenario in the VP-ESEM, due to the additional charge-control mechanisms provided by the gaseous environment, but is much more common in the high-vacuum SEM and is the classical example of charging

superimpose the decay characteristics onto this time interval, resulting in a steady charge build-up, as depicted in (c). This is the classical case of negative charging in the high-vacuum SEM. However, some subtle effects come into play in the VP-ESEM, and these will be discussed in Section 5.5.

5.4.2 Measuring Surface Potential

It is possible to determine whether a specimen has acquired a net positive or negative surface potential simply by examining an X-ray spectrum of the specimen (Newbury, 2002; Goldstein *et al.*, 2003). The idea is this: continuum X-rays (see Chapter 2, Section 2.4.4) will be generated by the primary electron beam up to some maximum energy: that of the primary beam itself. This is known as the Duane–Hunt limit.

If there is a positive potential at the surface, primary electrons will feel an attractive force and hence accelerate, and will land with a higher energy E_L than when they left the electron source. Continuum X-rays will therefore be generated up to some higher cut-off energy, thus giving a direct measure of the potential. Likewise, if there is a net negative potential on the specimen surface, primary electrons will feel a repulsive force and hence slow down, thereby landing with a lower energy E_L than when they started.

Experimentally, the value of E_L can be read from the energy axis of an X-ray spectrum as the point at which the continuum intensity falls to zero.[1] By comparing this value with the beam energy that was selected, it is easy to deduce the net gain or loss of energy experienced by the primary electrons and hence the surface potential V_s, according to:

$$V_s = (E_L - E_0)/e \qquad (5.5)$$

For example, if the user selects $E_0 = 20\,\text{keV}$ (by setting accelerating voltage $V_0 = 20\,\text{kV}$) and E_L is 'measured' to be $20.5\,\text{keV}$, then the surface potential $V_s = +500\,\text{V}$.

5.4.3 Conductive, Electrically Grounded Bulk Materials

Now, when primary electrons impinge on an electrically grounded material (e.g. a bulk metal), negative charges injected by the beam can

---

[1] Note that there may be a few counts beyond this point, due to pulse coincidence effects and dynamic charging and discharging, but these can safely be ignored for this purpose.

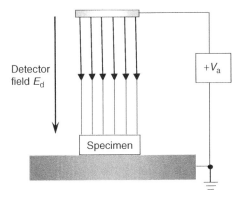

Figure 5.6   A simplified schematic diagram to show field lines between a positively biased anode and a grounded conductor. The field lines terminate on the specimen surface

flow freely to ground, via the specimen stage on which it is mounted. Indeed, any charge carriers, negative and positive, are free to move around and recombine to maintain charge neutrality.

In this simplest of cases, no electric field is set up within the specimen and the specimen is effectively at ground potential (zero volts). This means that the detector field lines terminate at the top surface of the specimen, as shown in Figure 5.6. The detector field strength E_d is a function of the anode bias V_a and anode–specimen distance d.

5.4.4 Conductive, Electrically Isolated Materials

Continuing with the theme of conductors for the moment, we now consider a bulk metal that has been separated from the specimen stage by an insulating material so that any excess charge carriers remain largely confined (i.e the charges cannot leak away), although they are still free to move around and redistribute themselves within the material.

Because of this ability for charges to move, the specimen cannot sustain an electric field. However, there is no direct path to ground for charge carriers to flow to or from. In this case, the excess charge manifests itself as an electric potential V_s at the surface. This is shown in Figure 5.7.

As with the grounded conductor discussed above, the detector field lines terminate on the surface of the specimen. However, the primary electron landing energy is affected by the specimen's electrical potential: reduced in the case of a negative potential and increased in the case of a positive potential, in accordance with Equation (5.2).

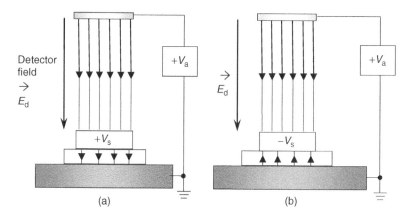

Figure 5.7 Schematic diagrams to show the electric potential that can build up around an electrically isolated conductive specimen during electron beam irradiation. Field lines still terminate at the specimen surface although in (a) a positive surface potential can accelerate primary electrons to the surface, increasing their landing energies, while in (b) a negative surface potential causes deceleration of the primary electron beam

In the high-vacuum SEM, a positive potential is associated with a total emission that exceeds unity, which occurs when the primary beam energy has a value between the first and second crossover energies, E_1 and E_2, respectively. In other words, the generation and escape of secondary electrons is so efficient between these energies that more electrons are emitted than the number entering the specimen.[2] Alternatively, beyond the E_2 point, the increase in primary electron penetration steadily reduces secondary electron emission efficiency, resulting in a net accumulation of primary electrons in the material and, hence, a negative potential arises. To some extent, this process is self-regulating. A demonstration of imaging in the presence of either positive or negative potentials in the VP-ESEM can be found in Toth *et al.* (2002c).

5.4.5 Nonconductive, Uncoated Materials

In the case of a nonconductive (i.e. electrically insulating) material, with no metallic coating to provide a ground plane at the surface, accumulated charges set up a dipole field within the specimen, rather like the situation shown in Figure 5.4 where electron emission leads

---

[2] This is, of course, a complex dynamic situation, but we will assume that we are dealing with the prevailing conditions of a given surface potential at a given moment.

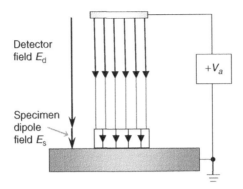

Figure 5.8 Schematic diagram to illustrate the electric fields in and around an electrically nonconductive, uncoated specimen during primary electron beam irradiation. The specimen can sustain an internal electric (dipole) field, and so the detector field lines pass through, to terminate at the grounded specimen stage

to a positive near-surface layer, separated from an underlying layer of negative charge due to the accumulation of thermalised primary electrons in the bulk.

An insulating specimen can sustain this electric field since charges neither rearrange nor leak away at an appreciable rate, as discussed in Section 5.4.1, and the internal dipole field (see Figure 5.4) means that the detector field lines extend through the specimen to terminate on the stage. This is shown in Figure 5.8. Primary electrons can therefore experience deceleration, decreasing their landing energies, further affecting beam penetration and the interaction volume, electron emission and charging, as already described.

Whilst schematics such as those shown provide a simple illustration of the effects of excess positive charges and implanted electrons, more sophisticated models of charge distribution and dipole fields have been extensively described by Cazaux (1985; 1999), for both the SEM and VP-ESEM. Similarly, an excellent review of the time-evolution of charging is given in Cazaux (2004).

Now, in the VP-ESEM we have, of course, an additional source of charge carriers: the positive ions created in the gas cascade process. These help to prevent the development of a negative surface potential such as that shown in Figure 5.7(b) or to mask the situation shown in Figure 5.8. But, under certain circumstances, these too can lead to the development of a positive surface potential and hence create a situation similar to that in Figure 5.7(a). In order to simplify matters for the moment, these effects have been omitted. However, we must eventually include this additional source of charge in our analysis, as it is a vitally important component. Therefore, the effects of positive ions will be dealt with separately, in Section 5.7, where we will look at their influence on surface potentials, dipole fields, detector field strength, primary electron landing energies and the emission or suppression of signals.

5.5 TIME- AND TEMPERATURE-DEPENDENT EFFECTS

5.5.1 Introduction

As outlined in Section 5.4.1, when the primary electron beam scans the surface of a specimen, the brief pulse of negative current at each point can result in the accumulation of negative charge in electrically nonconductive materials. However, the exact nature and lifetime of both the charge build up and its decay are dependent on a number of properties of the material. This can lead to a number of effects that we need to be aware of when interpreting images. In this section we will briefly survey some of the time- and temperature-dependent factors involved and how these affect the apparent electron emission from the specimen.

5.5.2 Conductivity and Some Time-Dependent Effects

The time constant for charge decay τ previously given in Equation (5.4) can be expressed in terms of the dielectric constant (i.e. polarisability) ε and conductivity σ (Blythe, 1980) of the material, such that:

$$\tau \propto \varepsilon/\sigma \qquad (5.6a)$$

where conductivity σ is given by the product of the number n of charges q and their mobility μ:, i.e.:

$$\sigma = nq\mu \qquad (5.6b)$$

Values of τ are said to be on the order of less than 10^{-20} seconds for copper and around 4×10^2 seconds for quartz (Thiel and Toth, 2005), whilst the time period to scan one frame in the SEM or VP-ESEM is typically between 10^{-1} and 10^2 s.

However, in order to appreciate the processes occurring within the bulk of an irradiated specimen, it is necessary to consider intrinsic and electron-beam-induced conductivity, as well as the leakage of charge occurring after the removal of the electron beam.

Intrinsic conductivity may arise via electronic or ionic mechanisms, depending upon the nature of the substance. For electronic conduction to occur, electrons must be free to move under the influence of an applied electric field (as occurs in metals, for instance). Ionic conduction involves the production of a current due to the movement of ions (positive and negative), particularly when there are a negligible number of electrons occupying states in the conduction band, as occurs in insulators (Azarov and Brophy, 1963).

Electron beam irradiation increases conductivity due to the creation of mobile, charged radicals and electronic defects. Electronic defects serve to trap charges, which may then be conducted by a 'hopping' mechanism, depending upon the energy and concentration of traps.

After removal of the electron beam, residual trapped charges are gradually released, so that delayed conductivity results. The rate of charge de-trapping depends on the energy (lifetime) of the traps. The distribution of trap depths also determines the ability to store charge (capacitance) – the deeper the trap, the longer a charge will remain in the trap. This capacitance gives rise to time-dependent conductivity, as a consequence of the rate of charge decay.

We can relate some of these ideas to the properties of various materials, from a good conductor to a perfect insulator, and their likely responses to the pulse of negative current from the primary electron beam as it dwells on a given point, as shown in Figure 5.9.

Firstly, as we would expect, a good conductor does not hold charge either during or after the pulse, while a material with slightly poorer conductive properties begins to conduct after a short build-up, and charge quickly drains away after the beam has moved on.

Clearly, the precise behaviour of a given material or region will vary as a result of intrinsic dielectric and conductive properties, localised defects and any additional induced and delayed conductivity.

For a more insulating material, some of the charge is held in the material and builds up, although some is conducted away during irradiation. When irradiation ceases on a given spot (i.e. as the primary beam continues its raster pattern), charge drains slowly, and provided that the electron beam does not revisit the same spot before the charge falls to zero, this point on the specimen will not accumulate charge over successive frames. Finally, for a perfect insulator, the accumulated charge remains in the material indefinitely and is not conducted away.

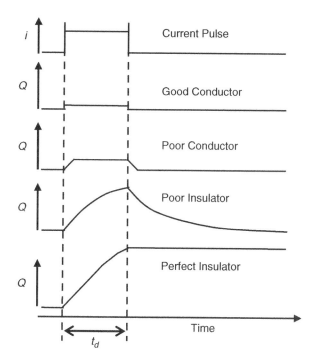

Figure 5.9 Diagrams to illustrate the properties of materials with varying degrees of insulating character and their response to a pulse of negative electrical current where i = current, Q = electrical charge build-up and t_d = dwell time of the primary electron beam. Reproduced with permission from Stokes *et al.* (2000). Copyright John Wiley and Sons, Inc

Return visits by the electron beam will simply increase the amount of accumulated charge (see Figure 5.5) until imaging becomes impossible.

However if we assume that, from one frame to the next, a specimen in the VP-ESEM is essentially electronically stable (negative and positive charge carriers in equilibrium), we find that it is still possible to induce a charge-related effect during the time that the electron beam is resident on a given point. In this case, the dwell time t_d becomes a key factor, as the charge accumulated can be sufficient to cause a transient localised enhancement in signal intensity (Stokes *et al.*, 2000). In fact, we can reformulate Equation (5.3) in terms of dwell time to find an expression for the charge σ_0 implanted per pixel N during irradiation (Equation (5.7)):

$$\sigma_0 = t_d I_0 (1 - \eta - \delta) \tag{5.7a}$$

where t_d is found by dividing the duration of the frame F by the product of the pixel dimensions N^2 (assuming a square display):

$$t_d = F/N^2 \tag{5.7b}$$

Of course, some charge will decay during this instant, in accordance with Equation (5.4), where $t = t_d$, and τ is of the order of t_d, such that σ_t is very sensitive to changes in dwell time, primary beam current and electron emission coefficient. Combining Equations (5.4) and (5.7) we have:

$$\sigma_t = t_d I_0 (1 - \eta - \delta)(e^{-t_d/\tau}) \tag{5.8}$$

By analogy with Figure 5.9, Figure 5.10 compares the properties of a poor conductor and a poor insulator in response to beam dwell times

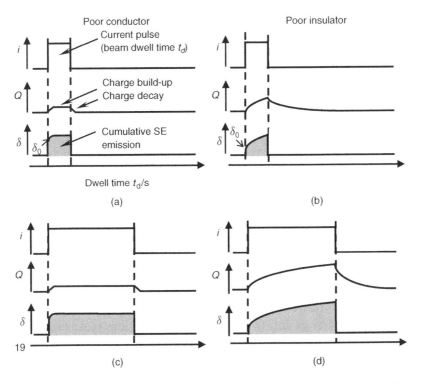

Figure 5.10 Schematic diagrams to indicate how different primary beam dwell times t_d can affect the accumulation of charge and hence secondary electron emission in materials with slightly different conductive properties. (a) and (b) are short dwell times, (c) and (d) longer dwell times. If these represent adjacent regions in the same specimen, the intrinsic secondary electron signal can become differentially increased to the extent that the contrast is reversed between these materials

of differing lengths and, in particular, the effect that these have on the apparent emission from each material. In this example, the poor conductor is assumed to have a higher secondary electron coefficient than the poor insulator and so in (a) the emission of secondary electrons is initially higher than for the material depicted by (b).

As charge starts to build up, increasing secondary electron emission, the cumulative signal collected during that instant continues to follow the trend that (a) appears more intense than (b). However, a situation can arise (i.e. long dwell time) that allows differences in charge accumulation to manifest themselves, in which case the cumulative secondary electron signal for the poor conductor is lower than that of the poor insulator, represented by (c) and (d), respectively. This causes a contrast reversal that can be readily returned to the intrinsic case by adjusting one or more of the parameters in Equation (5.8).

The effects of this transient contrast mechanism are demonstrated in Figure 5.11 for an oil-in-water emulsion. The less conductive oil phase (droplets) yield a higher secondary electron signal than the intrinsically brighter continuous water phase when the dwell time is increased. We will return to transient effects such as this and consider some example applications in Chapter 6.

Thus, in materials containing regions where charge can be trapped to differing degrees, it becomes possible to conceive of various scenarios

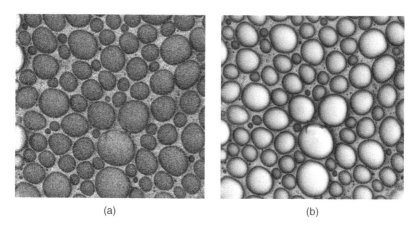

(a)   (b)

Figure 5.11 Instrinsic vs transient contrast in a liquid specimen consisting of an organic oil phase in a continuous water phase. (a) The oil phase intrinsically has a lower signal intensity than the water phase whereas, during higher electron flux (b), the intensity of the oil phase increases and the contrast is reversed. The ratio of dwell times between (a) and (b) is 1:60. Reproduced with permission from Stokes *et al.* (2000). Copyright John Wiley and Sons, Inc

where contrast can be maximised or minimised, depending on operating conditions (i.e. beam energy and local beam current, as determined by magnification and dwell time). A summary of these effects and references is given in Thiel and Toth (2005).

5.5.3 Charge Traps and Thermal Effects

We have noted already that a number of additional electronic states can arise within the energy gap that can trap excited electrons as they diffuse through a material (Section 5.3.1). These trapping states are typically due to defects and impurities, similar to the situation depicted in Figure 5.3, and the trap 'depths', as defined by the trapping energy, determine their lifetimes (i.e. the length of time an electron will remain immobilised). For further detail, descriptions of charge traps in the SEM and VP-ESEM are given in Toth *et al.* (2000), Thiel and Toth (2005) and Toth *et al.* (2007).

In order to demonstrate the influence of temperature on conductivity, Toth *et al.* (2000) visualised the effects of trapping, de-trapping and re-trapping of primary electrons caused by defects in an insulating material. Experiments were carried out using gallium nitride that had been pre-irradiated with high-energy helium ions to create defects in the bulk, giving regions with differing conductivities. Correspondingly, irradiation with a primary electron beam causes electrons to become trapped due to the defects in the less conductive region.

Raising the temperature of the specimen increases the energy of the trapped electrons to the point where they can escape the trap and leave the specimen. In this case, de-trapping occurs at temperatures in excess of about 300 °C. Subsequent irradiation as the temperature is reduced causes primary electrons to once again become trapped. These results are reproduced in Figure 5.12.

Note that the above work included an assessment of surface potential versus pressure, utilising the Duane–Hunt X-ray shift method outlined in Section 5.4.2. The negative surface potential due to accumulation of trapped electrons in the less conductive region reached several thousands of volts under high-vacuum conditions. The addition of nitrogen gas into the specimen chamber restored the Duane–Hunt limit, and hence primary electron landing energy E_L, to its correct value once the pressure reached $p = 40\,Pa$ (0.3 torr) under the specific conditions of the experiment.

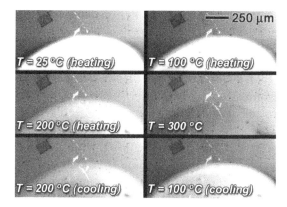

Figure 5.12 Primary beam electrons become caught in thermal charge traps in a specimen of gallium nitride, as shown by the bright region. When the temperature is raised, the electrons are released from their traps. Once the specimen begins to cool, electrons once more become trapped in the defects. Reproduced with permission from Toth *et al.* (2000). Copyright John Wiley and Sons, Inc

5.6 IMAGING SOFT MATERIALS

5.6.1 Introduction

In the high-vacuum SEM, specimens are frequently metal-coated, and so primary beam penetration depth is generally similar irrespective of the underlying material. Under these circumstances, the information content in an image does not particularly vary with different beam energies; hence it is quite common to find high-vacuum SEM work routinely carried out with a fixed primary beam energy, such as 20 keV, to ensure a well-focused beam spot and good signal-to-noise ratio.

However, imaging uncoated specimens in the VP-ESEM is quite a different matter and, in addition to the specific dielectric properties of the specimen, it is very important to take account of the effects of primary electron depth penetration and radiation damage. This is especially true for materials containing light elements, generally soft matter such as polymers and biological specimens where primary electron penetration can vary significantly with primary beam energy (see the discussion on stopping power in Chapter 2, Section 2.4.1.2). This brings both drawbacks and benefits, depending on circumstances.

5.6.2 Choosing an Appropriate Primary Beam Energy

To help visualise the range of primary electrons (as defined in Equation (5.1)) in an organic specimen, Figure 5.13 shows some Monte Carlo simulations for carbon at several different beam energies (1, 10 and 30 keV).

From these simulations we can infer that primary electrons travel up to ~20 nm into the specimen at an energy $E_0 = 1$ keV, Figure 5.13(a), with a lateral spread of around 20 nm from the centre of the interaction volume. Backscattered electrons are emitted from within the first 10 nm of the surface, with the majority coming from within ~5 nm.

For $E_0 = 10$ keV, Figure 5.13(b), the maximum primary electron penetration depth becomes around 1 µm, an increase of almost two orders of magnitude, consistent with Equation (5.1), with a corresponding lateral

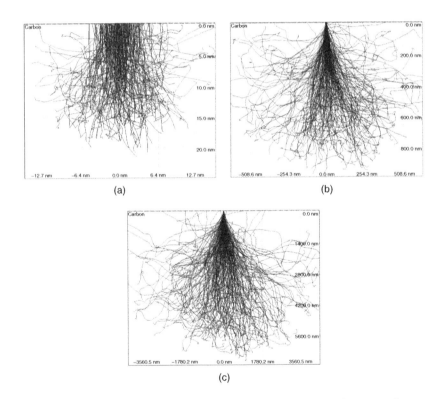

(a)   (b)

(c)

Figure 5.13 Monte Carlo simulations to show the penetration of primary electrons in carbon for beam energies $E_0 = 1$, 10 and 30 keV. In (a) the depth is marked every 5 nm. In (b) the markers are every 200 nm, while in (c) the depth markers occur every 1.4 µm. Some indicative values are given in the text. Simulated using CASINO v2.42 (Drouin *et al.*, 2007)

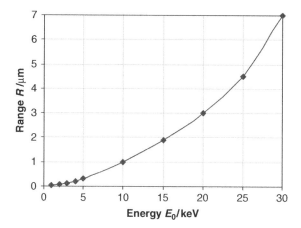

Figure 5.14 Plot to show the maximum penetration range R_{max} vs energy for primary electrons over a range of energies in carbon, estimated from Monte Carlo simulations

spread from the centre of $\sim 1\,\mu m$. The backscattered electron emission depth increases to within 400 nm of the surface.

In the case of primary beam energy $E_0 = 30\,keV$, Figure 5.13(c), the maximum depth is close to $7\,\mu m$, spreading out roughly $8\,\mu m$ laterally from the centre, with backscattered electrons emanating from around $2\,\mu m$ beneath the surface.

To give a better sense of perspective of the differences as a function of primary beam energy, the plot in Figure 5.14 summarises the trends in penetration depth for primary electrons in carbon.

5.6.2.1 Advantages of Low Primary Beam Energy

An example of the dramatic increase in primary electron range was shown in Chapter 2, Figure 2.20, where a thin polymer film could be rendered 'invisible' even at quite modest beam energies. Hence, for soft materials, it is advantageous to use a low primary beam energy in order to minimise beam penetration and maximise surface detail.

Furthermore, reducing the primary beam energy has a very significant impact on the generation of backscattered and backscatter-derived electron signals, due to consideration of the lateral primary electron range r_{max}. As shown in the previous section, reducing the size of the interaction volume reduces r_{max} and hence makes the lateral spatial resolution much better. Thus, even though the primary beam is more tightly

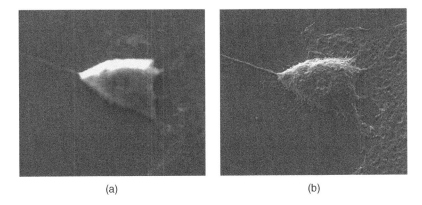

(a)   (b)

Figure 5.15 Uncoated organic specimens allow the primary electron beam to penetrate deeply. In (a) the beam energy is 12.5 keV and the details of this cell on a glass slide are difficult to distinguish. (b) Lowering the beam energy (3 keV in this case) gives a much better impression of the surface. Images obtained using water vapour, pressure $p = 546.6$ Pa (4.1 torr). Horizontal field width $= 50\,\mu$m. Images courtesy of Ellen Baken, FEI Company

focused at higher energies (see Chapter 2, Section 2.3), the advantage of this is lost for soft materials.

This is perfectly illustrated in Figures 5.15(a) and (b), where the spatial resolution is very poor for primary beam energy $E_0 = 12.5$ keV and much better for $E_0 = 3$ keV, where the primary electron penetration depth is much reduced. Using Figure 5.14 as a guide, primary electron penetration $R_{max} \sim 1.5\,\mu$m for $E_0 = 12.5$ keV, as opposed to $R_{max} \sim 100$ nm for $E_0 = 3$ keV.

5.6.2.2 Advantages of High Primary Beam Energy

Conversely, there are times when the characteristics described above can be used to our advantage to reveal, for example, depth information from an inorganic–organic specimen. Figure 5.16 consists of an organic emulsion (an off-the-shelf moisturising lotion) containing inorganic particles that are designed to impart specific properties (such as a shimmer effect when applied to the skin). Observations with different beam energies, $E_0 = 10$ keV vs $E_0 = 30$ keV in this case, help to provide depth information and an idea of the three-dimensional relationship between the particles and the surrounding medium.

Again using the information above and the data in Figure 5.14, primary electrons reach estimated depths of $1\,\mu$m for $E_0 = 10$ keV and $7\,\mu$m for $E_0 = 30$ keV.

(a)   (b)

Figure 5.16 Two micrographs of the exact same area of an emulsion consisting of an organic matrix with inclusions of inorganic particles. In (a) the primary electron penetration depth at $E_0 = 10\,$keV is relatively small compared to (b) where $E_0 = 30\,$keV, hence the inorganic particles can be seen at greater depths in (b). Imaged using water vapour, pressure $p = 400\,$Pa (3 torr)

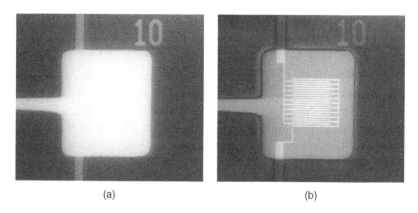

(a)   (b)

Figure 5.17 An optoelectronic transistor consisting of several layers, including polymers and gold, on a glass substrate. The beam energy in (a) is $E_0 = 15\,$keV, whilst in (b) $E_0 = 30\,$keV, enabling the buried source-drain region to be visualised. Horizontal field width = 2.4 mm. Specimen courtesy of Catherine Ramsdale, Plastic Logic Ltd

A second example, shown in Figure 5.17, is that of an optoelectronic transistor where the use of a higher beam energy reveals much more information about the underlying structure. At the base is a glass substrate, onto which an inter-digitated source-drain region has been laid down (evaporated gold tracks, ~50 nm thickness). Above this are several polymeric layers (~800 nm total thickness) and on the topmost surface is a gate electrode (evaporated gold, ~50 nm thickness).

At the lower energy, (a), the gate electrode is visible on the top surface, along with a buried numeral and part of a gold track running vertically down the image in areas where the primary beam only has to penetrate through polymer. At the higher beam energy, (b), the penetration is sufficient to reveal additional details below the gate electrode. In this case, the range is high enough to allow the beam to pass through the 50 nm thick gold region, allowing backscattered and backscatter-derived secondary electron signals to exit the top surface. Note that in the high-vacuum SEM specimen charging would make it impossible to exercise this degree of flexibility in beam energy, since the polymer and glass regions are highly insulating and the gold regions are electrically isolated.

5.6.3 Radiation Damage

Various feasibility studies have been carried out with regard to the handling and imaging of soft and/or hydrated materials, particularly those of a biological nature, to assess factors such as chamber environmental conditions (Cameron and Donald, 1994; Gilpin, 1997; Bache and Donald, 1998; Stokes, 2003) and radiation damage (Farley *et al.*, 1988; Jenkins and Donald, 1997; Kitching and Donald, 1998; Royall *et al.*, 2001). These have helped to increase awareness of some of the pitfalls to avoid when dealing with these delicate specimens.

Radiation damage can be a serious concern in general, as was outlined in Chapter 2, Section 2.5.1 for the case of high-vacuum SEM, but the situation is even more acute in the VP-ESEM, where samples are especially vulnerable for two main reasons:

1. samples have no coating to protect them from the electron beam;
2. the presence of water vapour as an imaging gas increases the formation of free radicals, known to be a major cause of radiation damage.

We saw in Chapter 2 that organic specimens may undergo a number of reactions as a result of radiation damage. Since many specimens observed in the VP-ESEM are of an organic nature, it is important to ensure that radiation damage is kept to a minimum, otherwise the interpretation of experiments may be undermined.

Factors that should be considered when imaging a sample include the following parameters of the microscope, as these may exacerbate radiation damage:

- accelerating voltage (or primary beam energy);
- magnification;
- scan rate;
- sample temperature;
- working distance;
- beam current density;
- chamber gas pressure.

The use of water vapour as an imaging gas results in the formation of free radicals and other charged species, in addition to radiolysis products created within the specimen itself, which can lead to rapid radiation damage.

Inelastic scattering results in the formation of ionised or excited water molecules, according to the schemes in Equations (5.9) and (5.10), respectively (Talmon, 1987):

$$e^- + H_2O \rightarrow H_2O^+ + 2e^- \tag{5.9}$$

$$e^- + H_2O \rightarrow H_2O^* + e^- \tag{5.10}$$

Water molecules in the excited state (Equation (5.10)) can then decay into free radicals (Equation (5.10a)) or ions (Equation (5.10b)):

$$H_2O^* + e^- \rightarrow H \cdot + OH \tag{5.10a}$$

$$H_2O^* + e^- \rightarrow H^+ + OH^- \tag{5.10b}$$

where H^+ is a proton and OH^- is the hydroxyl ion. Of these, the hydroxyl free radical OH is said to be the most abundant (Kitching and Donald, 1998; Royall et al., 2001).

Although comparatively little information has so far been established concerning radiation damage in the VP-ESEM, the available advice seems to be that reducing the primary beam energy helps to decrease the concentration of damaging species, as do short dwell times. For example, Royall et al. (2001) showed that for water vapour as the imaging gas, $E_0 = 5\,\text{keV}$ is preferable to 25 keV, and a dwell time $t_d = 10^{-4}$ s is better than $t_d = 10^{-3}$ s. It is clearly an essential prerequisite that, before any meaningful results can be obtained, each particular specimen is assessed and the most suitable imaging conditions determined, particularly if experiments are to be performed that involve hydration or dehydration of the specimen.

5.7 EFFECTS OF IONS ON IMAGING

5.7.1 Introduction

In this concluding section we turn our attention to the concentrations and behaviour of gaseous ions, the charge state of the specimen and the influence that these can have on each other under certain circumstances. Together with the information in this and the previous chapters, this will further help us to understand how and why certain imaging phenomena occur.

As we know, positive ions are generally a useful by-product of the gas ionisation cascade process, where they travel towards the sample surface and help to compensate against negative charge implanted in electrically nonconductive materials. Rather like the process given by Equation (5.9), ionisation can be described by the reaction scheme:

$$e^- + X \rightarrow X^+ + 2e^- \tag{5.11}$$

It is worth emphasising that the presence of these ions does not mean that negative specimen charging does not occur, but merely that the positive ions help to mitigate the effects of charging to enable successful imaging and microanalysis.

5.7.2 Consideration of the Concentration of Positive Ions

Now, since the number of ions generated is directly proportional to the number of electrons amplified, the gas cascade process may in fact lead to the production of many more positive ions than are actually needed for charge compensation. It was suggested that excess ions could interfere with the imaging process (Howie, 1997), and it was later shown that this is indeed the case: if the number (i.e. concentration) of ions is greater than the number needed to balance the negative charge in the specimen, this can have a number of effects on the system as a whole.

However, it can be rather difficult to choose conditions that satisfy the criteria that may be required, such as maintaining an adequately high pressure of gas for good signal amplification, especially for a given experiment (such as stabilising hydrated specimens at the appropriate water vapour pressure), while keeping the concentration of ions low enough to just compensate for negative charging. In addition, we must bear in mind that the gas amplification process is partially dependent on the electron emissive properties of the individual specimen or regions within it. Furthermore, there are effects related not only to the ions but to the dielectric properties of the specimen, and so these will need to be included in the discussions that follow.

5.7.3 Ion Mobility Effects

Due to their large mass, the drift velocity of ions is about three orders of magnitude lower than for electrons (Danilatos, 1990), so that ions appear almost stationary compared to the accelerating electrons. Under typical VP-ESEM conditions, the velocity v of ions is typically taken to be $v \sim 10^2 \, ms^{-1}$ and for electrons $v \sim 10^5 \, ms^{-1}$.

Given the nature of the cascade process, most of the ions are produced close to the anode that accelerates the electrons which is, of course, farthest from the specimen, as can be seen in Figure 5.18. The relatively slow drift rate of ions means that there is a finite transit time for the majority of ions to travel from this region to reach the specimen surface.

The slow response of the ions relative to electrons can lead to a lag between the onset of negative charging in the specimen and the arrival

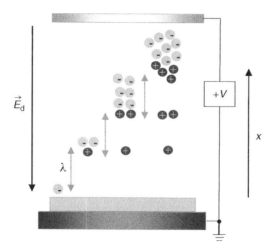

Figure 5.18 Schematic diagram to show the spatial distribution of electrons and ions during the cascade process. The majority of ions are produced near to the anode, although they are needed close to the specimen surface. A certain time-lag is therefore involved, which can be quite pronounced under circumstances of rapidly changing electron emission or other inhomogeneities in electrical properties. Adapted from a diagram originally presented by Matthew Phillips, University of Technology, Sydney

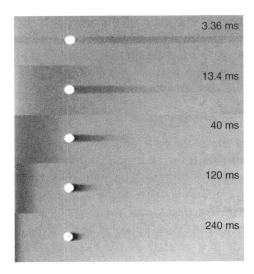

Figure 5.19 The slow drift rate of ions causes artefacts in images obtained via gas cascade amplification. The series shows how increasing the line scan time allows time for the system to react. Reproduced from Morgan and Phillips (2005), copyright Wiley-Blackwell

of positive ions, leading to an increase in signal intensity.[3] Likewise, if the beam moves to a region where fewer ions are needed, the signal becomes suppressed, due to the excess of ions. A symptom of these dynamic under- or over-compensation regimes is the presence of streaks that extend away from a feature in the x-direction, and whose magnitude changes in length as a function of scan rate (see Figure 5.19). Morgan and Phillips (2005) give an excellent overview of the various origins of ion-related imaging artefacts in the VP-ESEM, including those arising from the induced specimen current detection method. Furthermore, they have recently demonstrated a means to screen the slower part of ion drift, essentially eliminating smearing effects at fast scan rates (Morgan and Phillips, 2008).

The steady state concentration of ions is a property of secondary and backscattered electron emission coefficients (Chapter 3, Equation (3.2)), amongst other factors, and so if there is an abrupt change in emissive properties as the primary electron beam scans the surface of the specimen, this can cause the effect described above. Consider, for example, a sharp transition between regions of lower and higher

---

[3] This is also related to the detection method and the response time of the detector.

electron emission in a heterogeneous specimen. For the region of lower emissivity, the ion concentration is correspondingly lower, but when the emission increases on passing to the region of higher emissivity, there is a momentary increase in signal intensity until the ion concentration builds up. The effects are described in detail in Toth and Phillips (2000).

5.7.4 An Additional Surface Potential

5.7.4.1 Conductive, Electrically Isolated Materials

As discussed in Section 5.4, electrical potentials can develop in isolated conductors and nonconductive materials and these have an effect on both primary electron landing energies and secondary electron emission in the SEM and VP-ESEM. In Section 5.7.1 it was explained that in the VP-ESEM, an excess of positive ions can arise when the amplification process generates more ions than are needed for charge compensation.

One effect of these excess ions is that an additional potential can develop at the specimen surface since, aside from mutual repulsion, the ions do not dissipate. As a consequence, for cascade-dependent signals, this screens, and therefore reduces, the specimen–anode electric field strength which causes a lowering of the amplification efficiency, thus reducing the signal gain. The number of positive ions being produced is in turn reduced, of course, and hence there is an element of feedback: the concentration of positive ions goes down again and so the process is eventually self-limiting.

Experiments have been carried out by Craven *et al.* (2002) to show that for an initially charge-neutral specimen, a positive surface potential sets in and increases as the pressure of the imaging gas is raised. They used an electrically isolated bulk metal: a copper plate separated from the stage by an insulating layer, rather like the scheme shown in Figure 5.7(a). For a primary beam energy $E_0 = 20\,\text{keV}$, the measured surface potential rose to $V_s = +200\,\text{V}$ as the water vapour pressure increased from $p = 66.5\,\text{Pa}$ (0.5 torr) to $p = 665\,\text{Pa}$ (5 torr). Given that the potential on the detector anode was just $+400\,\text{V}$ in this experiment, this gives a substantial reduction in the detector field strength.

Incidentally, this brings us to an under-appreciated point regarding specimen preparation in the VP-ESEM. For the reasons just outlined, the use of insulating media such as double-sided adhesive tape to mount a specimen could easily result in problems even if the specimen is conductive, since the specimen can become isolated from ground potential just as in the case described above.[4] Hence, it is essential to always ensure a route to ground between the specimen and the stage and to verify the circuit by using a device such as a multi-meter.

5.7.4.2 *Nonconductive, Uncoated Materials*

An alternative scenario is that a net negative dipole field in a nonconductive specimen attracts positive ions to the surface but, since the charges move around less freely in an insulator, the charges remain above the surface. If the number of positive charges balances the negative charges, the net charge is zero.

However, in the situation where the ion concentration is high, ions build up in the gap between the specimen and the detector. This is commonly referred to as a cloud or layer of space charge, and is shown in Figure 5.20. This again results in a reduction in the detector field strength (Craven *et al.*, 2002).

5.7.5 Electron–Ion Recombination and Signal Scavenging

Another consequence of the potentials outlined in Section 5.7.4 is that secondary electrons emerging from the specimen can recombine with some of these excess positive charges. The removal of these potentially

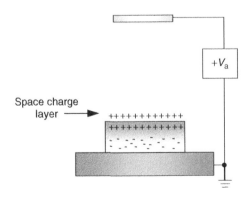

Figure 5.20 Schematic diagram depicting the layer of space charge in the specimen–anode gap. This acts to reduce the electric field strength that accelerates electrons towards the anode, reducing gas amplification and hence signal

---

[4] This method is absolutely fine for specimens mounted on stubs that are subsequently conductively coated for high-vacuum SEM but, in isolation, is unsuitable for VP-ESEM.

signal-forming electrons from the system so early in the process causes the signal to diminish (signal scavenging). The probability for electron–ion recombination decreases as the electrons gain energy in the electric field, and so secondary electrons that have moved away from the surface are less prone to being scavenged (Craven *et al.*, 2002; Toth *et al.*, 2002a; Toth *et al.*, 2002d; Morgan and Phillips, 2005).

On the other hand, if the mean free path of ions is long, either due to low ion concentration (e.g. pressures below ~100 Pa) or the electric field gradient is high, then positive ions can become particularly attracted to certain regions of the specimen, giving rise to preferential electron–ion recombination. This could occur in areas exhibiting pronounced electron emission associated with, for example, features such as edges or asperities. This effect is sometimes known as ion focusing (Toth *et al.*, 2002d). Again, signal scavenging occurs as a result, and can even lead to the inversion of contrast.

5.7.6 Combating Excess Ions

Clearly, if we can provide a means for preventing or controlling excess ions, we can avoid surface potentials and space charge, restoring the detector field strength and minimising signal scavenging.

In fact, there are several ways that could help to reduce the effects of positive ions. One way, suggested in Chapter 3, is to avoid the production of excess ions in the first place, perhaps by using the excitation–relaxation gas luminescence signal but remaining below the ionisation threshold (Section 3.3.3.2). This sounds promising but, as yet, there are insufficient experimental studies to support the idea. Alternatively, the use of an off-axis anode means that the majority of ions are generated in close proximity to the objective lens polepiece, and so have a ready-made path to ground. For good descriptions of this, see Toth *et al.* (2002b) and Thiel (2006).

Another way, known as a means for charge control within the high-vacuum SEM and focused ion beam (FIB) community, is to place some metallic tape along part of the surface of the specimen or close to a feature of interest, with the tape attached to the stage, ensuring a path to ground (e.g. by applying silver dag compound on the join between the tape and the stage). This works well for small areas, but is a rather crude approach, and the effects of the ions quickly come back into play as the distance from the tape increases (typically within a few tens to hundreds of micrometres).

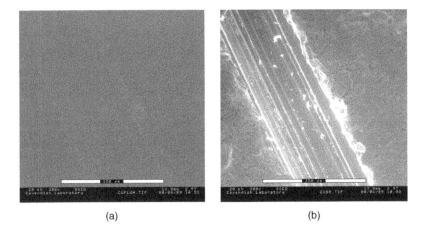

(a)	(b)

Figure 5.21 Images of the same area of a scratch on the surface of a copper specimen to demonstrate the detrimental effects of imaging a conductive, electrically isolated material that has built up a large positive surface potential (relative to the detector field E_d) in the VP-ESEM. In (a) the specimen is isolated from ground, allowing a positive surface potential to develop, which reduces the strength of the detector field while in (b) an array of grounded wires helps to give excess positive ions a path to ground so that they do not interfere with the imaging process. Reproduced from Craven *et al.* (2002), copyright Wiley-Blackwell

As an extension to this, a simple device was described by Craven *et al.* (2002), in which a series of wires, spaced 2 mm apart and positioned 1 mm above the specimen surface, served to collect excess ions and restore the detector field without interfering with it. The dramatic effects of this method are shown in Figure 5.21.

5.8 IMAGING WITH A GAS: SUMMARY

Taking into account all of the information in Chapters 3, 4 and 5, it is clear that there are numerous factors to consider when choosing an imaging gas, pressure, working distance and/or gas path length, and some of these depend on the electron emission characteristics of the specimen. Hence, the conditions required for a given specimen will vary on a case-by-case basis.

Ideas for reducing electron scattering and skirt radius were discussed throughout Chapter 4, and these included increasing the primary beam energy, using a chamber gas with low atomic number, using a low but adequate pressure or keeping the working distance short, but no closer than a distance equal to the diameter of the pressure-limiting (differential) aperture. It is worth re-stating that, for the purposes of X-ray microanalysis, the primary beam gas path length should be minimised, not by using a short working distance but, for example, by using some form of extension tube incorporating a pressure-limiting (differential) aperture (see Section 4.4.5) in order to maintain the take-off angle needed for X-rays to arrive at the detector.

The main points of imaging and microanalysis in a gas are summarised below, for ease of reference.

- Equation (4.2) describes the mean free path of primary electrons, taking into account elastic scattering probability, atomic number, molar mass and density of the gas
- Equation (4.8) tells us the useful primary electron beam current that remains, bearing in mind the current that has been lost due to scattering into the diffuse primary beam skirt
- Equation (4.9) gives us the parameters needed to characterise the radius of the primary electron beam skirt

These expressions suggest ways to minimise scattering, including:

- increasing the primary beam energy E_0;
- decreasing the atomic number Z of the gas;
- decreasing the pressure p;
- decreasing the gas path length GPL.

However, as we have seen in Chapters 4 and 5, these parameters are dependent on the requirements of a specific specimen or experiment.

Some things that can happen if the gas pressure is too low:

- gas amplified secondary electron signal gain is low;
- incomplete negative charge compensation can arise;
- positive ions can become focused onto specific areas of the specimen.

Some things that can happen if the gas pressure is too high:

- primary electron scattering increases;
- electron current in the focused beam decreases;
- positive ions can accumulate;
- detection of gas cascade electrons is reduced;

- X-ray signal intensities decrease;
- spurious X-rays can be detected from areas remote from the beam.

REFERENCES

Akkerman, A., Boutboul, T., Breskin, A., Chechik, R., Gibrechterman, A. and Lifshitz, Y. (1996). Inelastic electron interactions in the energy range 50eV–10KeV in insulators: Alkali halides and metal oxides, *Phys. Stat. Sol.. B*, **198**, 769.

Azarov, L.V. and Brophy, J.J. (1963). *Electronic Processes in Materials*. McGraw-Hill, Inc.

Bache, I.C. and Donald, A.M. (1998). The Structure of the Gluten Network in Dough: A Study Using Environmental Scanning Electron Microscopy. *J. Cereal Sci.*, **28**, 127–133.

Bishop, H. (1974). Electron–solid Interactions and Energy Dissipation, in *Quantitative Scanning Electron Microscopy*. D. Holt, M. Muir, P. Grant and I. Boswarva, Academic Press.

Blythe, A.R. (1980). *Electrical Properties of Polymers*, Cambridge University Press.

Cameron, R.E. and Donald, A.M. (1994). Minimising Sample Evaporation in the Environmental Scanning Electron Microscope. *J. Microsc.*, **173**(3), 227–237.

Castell, M.R., Perovic, D.D. and Lafontaine, H. (1997). Electronic contribution to secondary electron compositional contrast in the scanning electron microscope. *Ultramicroscopy*, **69**(4), 279–287.

Cazaux, J. (1985). Some Considerations on the Electric Field Induced in Insulators by Electron Bombardment. *J. Appl. Phys.*, **59**(5), 1418–1430.

Cazaux, J. (1999). Some Considerations on the Secondary Electron Emission δ, from e^- Irradiated Insulators. *J. Appl. Phys.*, **85**(2), 1137–1147.

Cazaux, J. (2004). About the mechanisms of charging in EPMA, SEM and ESEM and their time evolution. *Microsc. Microanal.*, **10**, 670–684.

Chi, J.Y. and Gatos, H.C. (1979). Determination Of Dopant-Concentration Diffusion Length And Lifetime Variations In Silicon By Scanning Electron-Microscopy. *J. Appl. Phys.*, **50**(5), 3433–3440.

Craven, J.P., Baker, F.S., Thiel, B.L. and Donald, A.M. (2002). Consequences of Positive Ions upon Imaging in Low vacuum SEM. *J. Microsc.*, **205**(1), 96–105.

Danilatos, G.D. (1990). Theory of the Gaseous Detector Device in the Environmental Scanning Electron Microscope. *Adv. Electron. Electr. Phys.*, **78**.

Doehne, E. (1998). Charge Contrast: Some ESEM Observations of a New/Old Phenomenon. *Microsc. Microanal.*, **4**(Suppl.2: Proceedings), 292–293.

Drouin, D., Couture, A.R., Joly, D., Tastet, X. and Aimez, V. (2007). CASINO V2.42 – A fast and easy-to-use modeling tool for scanning electron microscopy and microanalysis users. *Scanning*, **29**, 92–101.

Elliott, S.L., Broom, R.F. and Humphreys, C.J. (2002). Dopant profiling with the scanning electron microscope – A study of Si. *J. Appl. Phys.*, **91**(11), 9116–9122.

Farley, A.N., Beckett, A. and Shah, J.S. (1988). Meatsem (Moist Environment Ambient-Temperature Scanning Electron Microscopy) Or The SEM Of Plant Materials Without Water Loss. *Inst. Phys. Conference Series*, **93**, 111–112.

Gilpin, C.J. (1997). Biological Applications of Environmental Scanning Electron Microscopy. *Microsc. Microanal.*, **3**(2), 1205–1206.

Goldstein, J., Newbury D, Joy, D, Lyman, C, Echlin, P, Lifshin, E, Sawyer, L and Michael, J (2003). *Scanning Electron Microscopy and X-Ray Microanalysis*, third edition, Plenum.

Griffin, B.J. (2000). Charge Contrast Imaging of Material Growth Defects in Environmental Scanning Electron Microscopy – Linking Electron Emission and Cathodoluminescence. *Scanning*, **22**, 234–242.

Howie, A. (1995). Recent Developments in Secondary Electron Imaging. *J. Microsc.*, **180**(3), 192–203.

Howie, A. (1997). Personal communication.

Jenkins, L.M. and Donald, A.M. (1997). Use of the Environmental Scanning Electron Microscope for the Observation of the Swelling Behaviour of Cellulosic Fibres. *Scanning*, **19**, 92–97.

Kanaya, K. and Okayama, S. (1972). Penetration And Energy-Loss Theory Of Electrons In Solid Targets. *J. Phys. D – Appl. Phys.*, **5**(1), 43–58.

Kitching, S. and Donald, A.M. (1998). Beam Damage of Polypropylene in the Environmental Scanning Electron Microscope: an FTIR Study. *J. Microsc.*, **190**, 357–365.

Kittel, C. (1986). *Introduction to Solid State Physics*, John Wiley & Sons, Inc.

Lide, D.R. (Ed.) (1991). *CRC Handbook of Chemistry and Physics*, CRC Press.

Morgan, S.W. and Phillips, M.R. (2005). Transient analysis of gaseous electron–ion recombination in the environmental scanning electron microscope. *J. Microsc.*, **221**(pt 3), 183–202.

Morgan, S.W. and Phillips, M.R. (2008). High bandwidth secondary electron detection in variable pressure scanning electron microscopy using a Frisch grid. *J. Phys. D – Appl. Phys.*, **41**(5).

Newbury, D.E. (2002). X-ray microanalysis in the variable pressure (environmental) scanning electron microscope. *J. Natl Inst. Standards Technol.*, **107**(6), 567–603.

Royall, C.P., Thiel, B.L. and Donald, A.M. (2001). Radiation Damage of Water in Environmental Scanning Electron Microscopy. *J. Microsc.*, **204**(pt 3), 185–195.

Seiler, H. (1983). Secondary Electron Emission in the Scanning Electron Microscope. *J. Appl. Phys.*, **54**(11).

Shaffner, T.J. and Van Veld, R.D. (1971). 'Charging' Effects in the Scanning Electron Microscope. *J. Phys. E – Sci. Instr.*, **4**, 633–637.

Stokes, D.J. (2003). Recent advances in electron imaging, image interpretation and applications: environmental scanning electron microscopy. *Phil. Trans. Roy. Soc. London Series A – Math. Phys. Eng. Sci.*, **361**(1813), 2771–2787.

Stokes, D.J., Thiel, B.L. and Donald, A.M. (1998). Direct Observations of Water/Oil Emulsion Systems in the Liquid State by Environmental Scanning Electron Microscopy. *Langmuir*, **14**(16), 4402–4408.

Stokes, D.J., Thiel, B.L. and Donald, A.M. (2000). Dynamic Secondary Electron Contrast Effects in Liquid Systems Studied by Environmental SEM (ESEM). *Scanning*, **22**(6), 357–365.

Talmon, Y. (1987). Electron Beam Radiation Damage to Organic and Biological Cryo-specimens, in *Cryotechniques in Biological Electron Microscopy*, RA Steinbrecht and K. Zerold., Springer-Verlag.

Thiel, B.L. (2006). Imaging and microanalysis in environmental scanning electron microscopy. *Microchim. Acta*, **155**(1–2), 39–44.

Thiel, B.L. and Toth, M. (2005). Secondary electron contrast in low-vacuum/ environmental scanning electron microscopy of dielectrics. *J. Appl. Phys.*, **97**(5).

Toth, M., Daniels, D.R., Thiel, B.L. and Donald, A.M. (2002a). Quantification of electron–ion recombination in an electron-beam-irradiated gas capacitor. *J. Phys. D – Appl. Phys.*, **35**(14), 1796–1804.

Toth, M., Knowles, W.R. and Phillips, M.R. (2007). Imaging deep trap distributions by low vacuum scanning electron microscopy. *Appl. Phys. Lett.*, **90**(7).

Toth, M., Kucheyev, S.O., Williams, J.S., Jagadish, C., Phillips, M.R. and Li, G. (2000). Imaging charge trap distributions in GaN using environmental scanning electron microscopy. *Appl. Phys. Lett.*, **77**(9), 1342.

Toth, M. and Phillips, M.R. (2000). The role of induced contrast on images obtained using the environmental scanning electron microscope. *Scanning*, **22**, 370–379.

Toth, M., Phillips, M.R., Craven, J.P., Thiel, B.L. and Donald, A.M. (2002b). Electric Fields Produced by Electron Irradiation of Insulators in a Low Vacuum Environment. *J. Appl. Phys.*, **91**(7), 4492–4499.

Toth, M., Phillips, M.R., Thiel, B.L. and Donald, A.M. (2002c). Electron Imaging of Dielectrics under Simultaneous Electron–Ion Irradiation. *J. Appl. Phys.*, **91**(7), 4479–4491.

Toth, M., Thiel, B.L. and Donald, A.M. (2002d). On the Role of Electron–Ion Recombination in Low Vacuum SEM. *J. Microsc.*, **205**(1), 86–95.

6

A Lab in a Chamber – *in situ* Methods in VP-ESEM and Other Applications

6.1 INTRODUCTION

This concluding chapter has several aims, one of which is to highlight certain aspects of VP-ESEM where, by conventional high-vacuum SEM standards, the information that can be obtained is unusual and in many cases truly unique to VP-ESEM. The examples here serve to demonstrate how aspects such as the presence of a gas, the absence of an electrically conductive coating and control over the physical state of the specimen can be utilised in different ways to give new insights, as well as results that are complementary to other techniques. The examples collected here may even help to stimulate ideas for new and/or improved uses of the VP-ESEM, continuing its evolution.

The chapter is divided into three areas. The first explores the specialised information that is available in the VP-ESEM and hence ways for imaging nonconductive materials at the nanoscale in the absence of a coating. Then, since the VP-ESEM naturally lends itself to dynamic *in situ* experiments, the various techniques developed so far will be reviewed. Amongst the work discussed will be mechanical testing, experiments at high and low temperatures and utilisation of gas chemistry for nanolithography and nanofabrication. The final section acts as a survey of the literature, touching on a variety of other subjects involving static imaging.

Principles and Practice of Variable Pressure/Environmental Scanning Electron Microscopy (VP-ESEM)
D. J. Stokes
© 2008 John Wiley & Sons, Ltd

Throughout this chapter we will deal with a wide range of applications involving soft, hard and composite materials, in multidisciplinary fields spanning semiconducting materials and devices to life science specimens and applications in art, conservation and forensics.

6.2 NANOCHARACTERISATION OF INSULATING MATERIALS

6.2.1 High-Resolution Imaging

Imaging of nanoscale features requires a highly focused primary beam and high magnifications. However, for an uncoated, bulk insulating material, this is a real challenge, even in the VP-ESEM.

Several recent publications have shown that it is possible to image highly insulating nanostructured materials with high resolution, in the absence of charging artefacts. For example, Toth *et al.* (2006) imaged chromium features on a thick quartz photolithographic mask, a very difficult specimen to image in an SEM, and demonstrated that resolution on the order of a nanometre or so can be achieved, using relatively low gas pressures (40–100 Pa, 0.3–0.75 torr) and maximising the number of ionising collisions along the path of secondary electrons using electric and magnetic fields.

The fields cause low-energy (secondary) electrons to follow a circular path whilst rotating in a magnetron motion, thus greatly increasing the effective path length for ionising collisions while at the same time keeping the working distance small to keep the specimen within the field of a magnetic immersion lens, improving resolution. The motion of the accelerated secondary electrons is demonstrated in Figure 6.1. The final component is a built-in trap to ensure proper control over excess ions, thus ensuring high signal gain without the risk of the difficulties posed by having an excess of ions. Further details can be found in, for example, Thiel *et al.* (2006).

This imaging method was also used by Kucheyev *et al.* (2007) to study aerogels: ultra-low-density silica, with a highly open nanoporous structure, commonly used as a collecting medium for micrometeorites and other space-borne particles. In the bulk, they are extremely fragile and difficult to prepare, as well as being highly insulating and so difficult to image in the SEM. However, Kucheyev and co-workers were successfully able to investigate a range of morphologies and pore sizes of the bulk material in the VP-ESEM.

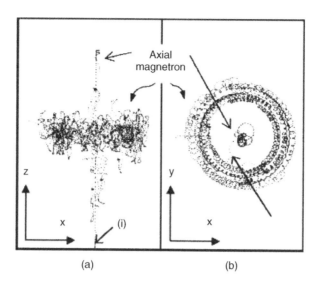

Figure 6.1 Monte Carlo simulations to show the path of just one secondary electron passing through electric and magnetic fields in the VP-ESEM. The side view (a) shows how the electron progresses in the z-direction in the electric field until it reaches the plane of the magnetic field where it begins magnetron motion in the $x-y$-direction. Eventually it emerges above the plane, where it once more accelerates in the electric field, towards the anode. (b) Plan view showing the circular paths of the electron. Reproduced from Thiel *et al.* (2006), copyright American Institute of Physics

Figure 6.2 shows a nanostructured oxide material, imaged in the manner described above, at a relatively low primary beam energy (5 keV) and low pressure (40 Pa, 0.3 torr).

6.2.2 Anti-Contamination in the VP-ESEM

Using the SEM for the measurement of critical dimensions for bulk materials at the nanoscale is of great importance, particularly in the semiconductor industry (see Section 6.2.3), but the results can be affected not just by charging artefacts but also by contamination: carbonaceous material that builds up on the specimen surface at the primary beam impact point, obscuring fine detail. This is a well-known problem in the high-vacuum SEM.

Sources of contamination include back-streamed vapour from vacuum pump oil and dirt on the chamber walls, but often it is the specimen itself that either has sources of contamination on its surface or contains components that contribute to the liberation and adsorption of carbon-containing deposits.

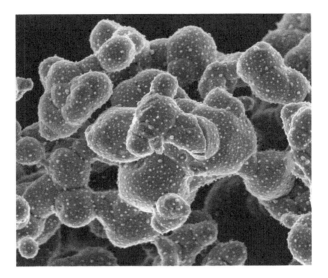

Figure 6.2 Silver oxide ceramic imaged using water vapour with pressure $p = 40$ Pa (0.3 torr). Primary beam energy $E_0 = 5\,keV$. Horizontal field width $= 11\,\mu m$. Image courtesy of Paul Gunning, Smith and Nephew Technical Services Group

However, in the VP-ESEM, contamination is significantly reduced or eliminated, and even if a specimen has been exposed to contamination in high-vacuum SEM, subsequent imaging in VP-ESEM can remove the contamination (see Figure 6.3). Toth *et al.* (2005) have proposed a mechanism whereby adsorbed contaminants are preferentially desorbed from the specimen surface by the combined interaction of primary electrons with the imaging gas and the surface. This is an excellent example of the additional power of the VP-ESEM for obtaining high-resolution information for any material, conductive or insulating.

6.2.3 Nanometrology

For uncoated electrically insulating materials, then, the combination of contamination- and charge-free imaging is very powerful indeed, and Postek and Vladar (2004) have outlined the advantages of this approach for semiconductor inspection and metrology. In the high-vacuum SEM, inspection of masks and wafers and the measurement of critical dimensions generally requires low voltages together with an understanding of any inherent charging of the specimen that may cause deflection of the primary electron beam. This type of work is therefore often accompanied by modelling to assist with accuracy.

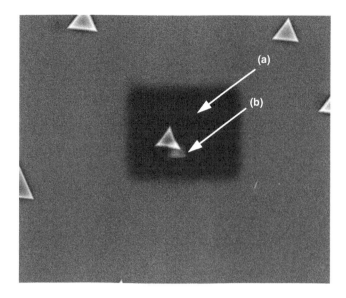

Figure 6.3 Secondary electron image demonstrating (a) the presence of contamination due to imaging in high vacuum (large dark rectangle) and (b) its removal in VP-ESEM mode (small bright rectangle). Image courtesy of Milos Toth, FEI Company

Postek and Vladar show how VP-ESEM can be applied with ease for viewing photomasks and photoresists, with high resolution and without the need for charge-modelling. In the semiconductor industry, control over ever-decreasing line widths and subsequently the accurate measurement of critical dimensions is a crucial step in the silicon wafer manufacturing process.

Figure 6.4 shows part of a chromium detail on a quartz photomask, where the roughness of the line has been measured down to a few nanometres. The quartz substrate is highly nonconductive and the chromium, although electrically conductive, is isolated from ground by virtue of being supported on the 150 mm thick insulating quartz layer, rather like the situation discussed in Chapter 5, Section 5.4.4.

6.2.4 Utilising Novel Contrast Mechanisms

We saw in Chapter 5, Section 5.5 that charge trapped in a specimen can affect secondary electron emission in such a way that it gives rise to subtle contrast effects. This 'stable' charging in the gaseous environment of the VP-ESEM provides the opportunity to utilise some unique contrast

Linewidth: 167 nm
Left edge roughness: 5.7 nm
Right edge roughness: 6.1 nm

Figure 6.4 High-resolution VP-ESEM secondary electron image of a chromium line on a quartz photomask showing roughness measurements of around 6 nm along both edges. Imaged with water vapour pressure $p = 100$ Pa (0.75 torr), primary beam energy $E_0 = 9$ keV. Horizontal field width = 900 nm. Reproduced from Postek and Vladar (2004), copyright John Wiley and Sons, Inc

mechanisms, effectively revealing the distribution of impurities and defects (charge traps), giving access to valuable information about the specimen that might not otherwise be available.

For example, Clausen and Bilde-Sorensen (1992) and Horsewell and Clausen (1994) showed the effects of varying primary beam energy, current, scan rate and pressure (ion concentration) on images of an yttria-doped zirconia ceramic. Two effects were reported: a strong enhancement in the definition of grain boundaries (Clausen and Bilde-Sorensen, 1992) and the presence or absence of structural information depending on the proximity of a path to ground (such as a more conductive adjacent area) with details fading as the ground plane is approached (Horsewell and Clausen, 1994). (See also Figure 6.5 for a similar observation).

Likewise, other workers have demonstrated these potentially useful effects in a range of nonconductive mineral-based materials. Examples include polycrystalline diamond (Harker *et al.*, 1994) and Travertine limestone (Doehne, 1998). The latter material predominantly consists of calcite ($CaCO_3$), with minor and trace elements of various metals, but yet exhibits stark contrast under certain conditions. Similar results were reported by Griffin (1997; 2000) and Baroni *et al.* (2000) when studying synthetically grown crystals of the mineral gibbsite [$Al(OH)_3$], an intermediate phase used in the Bayer process to convert bauxite ore into alumina (Al_2O_3). It was shown that there is a correlation between impurities (calcium and iron) incorporated into the solution-grown crystal structure, and image contrast due to charge traps representing the growth zones. Indeed, these zones could be matched to the number of times that a specific crystal had been recirculated through the batch precipitation process (Baroni *et al.*, 2000). The induced contrast was later used to model a three-dimensional representation of impurity distribution in gibbsite (Baroni *et al.*, 2002).

Figure 6.5 demonstrates the vivid contrast that can be induced, again dependent on the operating conditions mentioned above. In this case, pressure $p \sim 150$ Pa (1.1 torr) and primary beam energy $E_0 = 30$ keV. On the right-hand side of each image, impurity-related growth zones are clearly visible across the uncoated gibbsite grains. On the left-hand side, grains are beneath a thin conductive carbon coating and there is no contrast: the grains exhibit uniform signal intensity as the electrical

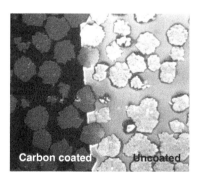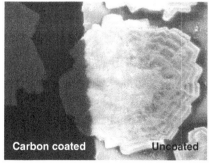

Figure 6.5 Low and high magnification images to show charge-related contrast in the mineral gibbsite $Al(OH)_3$. Imaged in water vapour with pressure $p \sim 150$ Pa (1.1 torr). Primary beam energy $E_0 = 30$ keV. Growth zones are revealed when the parameters are adjusted to allow defects to be seen (right-hand side of each image). A thin layer of carbon coating immediately suppresses the effect (left-hand side of each image). Horizontal field widths $= 2.2$ mm and $200\,\mu$m, respectively. Images courtesy of Brendan Griffin, University of Western Australia

characteristics of the system are altered. Notice how the contrast starts to fade out as the conductive coating is approached.

Note that, in all of the cases described, the specimens were flat-polished to eliminate surface topography, and so this can be reliably discounted as the source of the observed effects. Furthermore, surface scratches and debris are found to diminish any induced contrast.

In some cases, the contrast is more pronounced if charge is first allowed to accumulate in the specimen for a short time while in other cases this would cause a reduction in contrast. In many cases the effect is induced when the gas pressure is low and the anode bias high. These are similar to the conditions outlined in Chapter 5, Section 5.7.5. However, the interdependency of parameters means that there are a number of different ways to induce contrast in the VP-ESEM. For example, Craven *et al.* (2002) showed that under conditions of high ion flux ($\sim$530 Pa, 4 torr) the induced gibbsite structure is absent but, when the excess ions are provided with a path to ground (see Chapter 5, Section 5.7.6), the pronounced characteristic contrast is again revealed. The correlation between induced contrast and the presence of defects, particularly for gibbsite, has also been studied using cathodoluminescence (Chapter 2, Section 2.4.5), which is sensitive to the electronic structure of insulators and semiconductors and can give information about the size of the energy gap and any impurity-related deviations (Griffin, 2000). In addition, surface potentials of insulators can be studied using Kelvin probe microscopy (Kalceff, 2002).

Several studies on gibbsite can be found in the literature,[1] and an excellent starting point for the interested reader is a special issue of *Microscopy and Microanalysis* (Multi-authors, 2004) on the subject of characterising nonconductive materials, following a congress that took place prior to the Microscopy and Microanalysis 2002 meeting.

In other studies, Phillips *et al.* (1999) demonstrated scan-rate-dependent induced contrast across the *p*–*n* junction of a silicon diode in the VP-ESEM, showing depletion layers and suggesting the possibility for analysing metal-oxide-semiconductor devices. The subject is discussed further in Toth and Phillips (2000). As we saw in Chapter 5, Section 5.5.2, a transient contrast effect was similarly reported for a liquid-state heterogeneous specimen (Stokes *et al.*, 2000).

Zhu and Cao (1997) have studied negative and positive ferroelectric domains in lithium tantalum oxide using the VP-ESEM, noting

---

[1] It should be noted that there are numerous abstracts on this subject in conference proceedings, especially those of Microscopy and Microanalysis, that are not listed here.

that, again, highly polished specimens are needed along with high primary beam current and relatively low pressures, and Xiao *et al.* (2002) obtained images of domains in lead titanates. In addition, Pooley (2004) has described how induced contrast provides valuable chemical information on mineral distribution in chromian spinel and Cuthbert and Buckman (2005) have shown how this method can reveal growth zones in garnet.

The recent work of Clode (2006) is particularly striking, and the potential for charge-related contrast in biological specimens is very nicely demonstrated. As can be seen in Figure 6.6(a), an element of controlled charging reveals structure in the secondary electron image that would otherwise be hidden. A comparison of Figures 6.6(a) and (c) shows that when a conductive coating is applied, as in (c), the

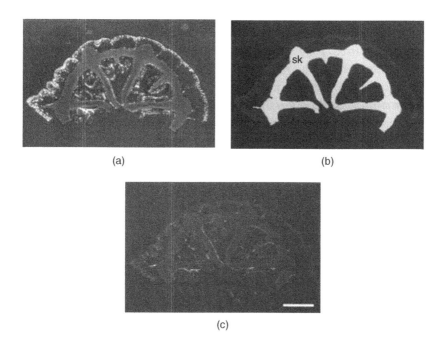

(a)   (b)

(c)

Figure 6.6   Images to demonstrate charge-related contrast in a biological specimen (coral polyp). (a) Electrons trapped in the material give rise to stable charging and hence contrast, particularly in the soft tissue regions. Compare this with a backscattered electron image, (b), which vividly shows regions containing calcium and (c) a secondary electron image after the application of a carbon coating. Imaged with beam energy $E_0 = 30\,keV$ and water vapour pressure $p = 27\,Pa$ (0.2 torr). Horizontal field width = 3 mm. Reproduced from Clode (2006), copyright Elsevier

effect is lost since the coating provides a termination point for electric fields present in or above the specimen. Figure 6.6(b) is a backscattered electron image that picks out the central mineral part of the organism as distinct from the low atomic number soft tissue. This series of images clearly shows that the charge-related secondary electron information in (a) is a unique way to visualise ultra-structural detail.

Many of the reports in the literature arise from imaging via the gas cascade amplification mechanism. However, detection via the induced specimen current method has been shown to produce similar effects for polymeric materials (Gauvin *et al.*, 2003).

6.2.5 Transmitted Electron Signals – STEM and wetSTEM

In addition to the more familiar backscattered and secondary electron detectors, positioning of a solid-state detector beneath a thin specimen in the SEM allows for the collection of primary electrons that have been transmitted right through the sample. In order to distinguish this technique from the more powerful scanning transmission electron microscope (STEM), this form is usually known as STEM-in-SEM (or 'poor man's STEM').

Some of the primary electrons will interact more strongly with the specimen than others as they travel through the specimen, because of differences in atomic number, density or thickness, and this will correspondingly cause the electrons to emerge at a range of different scattering angles. This principle is illustrated schematically in Figure 6.7.

A typical STEM-in-SEM detector has different segments that can be used to distinguish between electrons arriving at these various angles. For example, electrons that are hardly scattered at all will arrive in the straight-through direction to form a bright field, BF, signal, and those that have scattered to higher angles a dark field, DF, signal. This produces contrast between heterogeneities in the specimen.

An extension to the STEM-in-SEM idea is to combine it with the use of a water vapour environment to control the temperature of the specimen, as described in Chapter 3, Section 3.4, to enable thin wet specimens to be viewed in STEM mode. This can be achieved by Peltier-cooling the TEM grid on which the specimen is held, as shown schematically in Figure 6.8, and utilising the water vapour capabilities of the microscope.

Figure 6.9 shows a specimen that has been prepared for the TEM but imaged under hydrating conditions in the VP-ESEM using an arrangement similar to that shown in Figure 6.8.

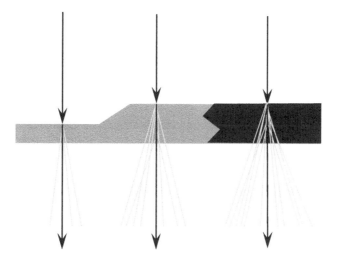

Figure 6.7 Transmission of primary electrons through a thin specimen. In regions where the specimen is thinner (on the left), only a few primary electrons will be scattered, relative to a thicker region of the same density (centre). In regions of similar thickness but different atomic number Z, electrons will scatter more strongly through regions of higher Z, as indicated by the darker region to the right. Adapted from Goodhew *et al.* (2001)

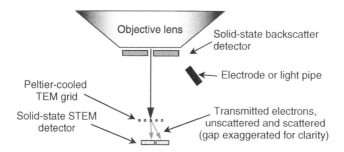

Figure 6.8 Simplified schematic diagram to show the concept of wetSTEM in the VP-ESEM. A Peltier-cooled TEM holds a thin sample, which can contain aqueous phases by using the appropriate specimen temperature and pressure of water vapour, and a solid-state detector collects the transmitted electrons to form high-resolution bright field, dark field and, depending on the detector, annular dark field STEM images

Note that beam energies in the SEM, up to 30 keV, are quite a bit lower than in the conventional STEM instrument which can operate at energies as high as 300 keV. Now, a particular difficulty in observing organic materials in the STEM is that they have inherently low atomic

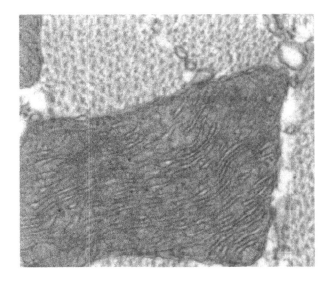

Figure 6.9 STEM-in-SEM image of a thin section (100 nm thickness) of heart tissue, fixed and stained for the TEM using uranyl acetate and lead citrate. Although this particular specimen is not hydrated, the chamber conditions mimic those required to maintain a hydrated specimen: water vapour pressure $p \sim 600$ Pa (4.5 torr), temperature $T = 2\,°C$. Horizontal field width $= 1.25\,\mu m$. Courtesy of Chris Gilpin, University of Texas

numbers Z, so their electron-stopping powers are very low. As we have seen throughout, higher energy electrons interact less strongly with matter: they scatter less, and so organic materials produce lower contrast in high-energy STEM.

However, when employing a STEM detector in the SEM or VP-ESEM with their lower beam energies, it becomes easier to differentiate between similar, low-Z materials and so we can form high-contrast STEM images of organic materials without necessarily using heavy metal salts to stain the different components. Williams *et al.* (2005) demonstrated this for semiconducting polymer blends and Bogner *et al.* (2005; 2007) showed how this method can be used to study a range of specimens such as emulsions and nanoparticles in suspension, with no staining. An example is shown in Figure 6.10. Similarly, Doehne and Baken (2006) have shown the application of wetSTEM to conservationally important specimens such as lime mortars and pigment fibres. In the latter case, they observed ultra-structural information in greater detail than that previously shown by SEM alone. This type of work in the VP-ESEM is still at an early stage, but is a promising direction for the future.

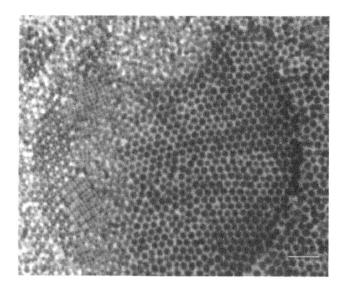

Figure 6.10   wetSTEM image of acrylic latex particles in suspension. Horizontal field width = 5 μm. Reproduced from Bogner *et al.* (2005), copyright Elsevier

6.3 *IN SITU* EXPERIMENTS

6.3.1 Deformation and Failure

Various types of tensile stage can be used in the VP-ESEM. This allows tensile or compressive mechanical testing and analysis of specimens with the advantage, of course, that specimens are uncoated. Furthermore, the load versus extension data can be collected simultaneously, which can be directly related to stress–strain behaviour and *in situ* microscopic observations. These could include brittle-to-ductile transitions, elastic and plastic failure and so on. Examples of *in situ* deformation experiments of nonconductive materials include: crack growth at the porous interfaces of ceramic layers (Sorensen *et al.*, 2002); a rubber-toughened film (He and Donald, 1997); and maltodextrin particles in a biopolymer matrix (Rizzieri *et al.*, 2003a; 2003b). Figure 6.11 shows the effect of *in situ* tensile testing on a maltodextrin particle in a gelatin gel matrix.

In addition, the tensile stage can be cooled to allow moisture-containing materials to be tested in the hydrated state. Examples of this approach include the deformation and failure of parenchymal cells of carrot (Thiel and Donald, 1998; Warner *et al.*, 2000), biopolymer

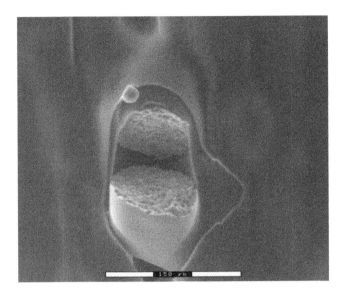

Figure 6.11   Fracture failure of a particle embedded in a polymer during an *in situ* tensile experiment in the VP-ESEM. Water vapour pressure $p \sim 640$ Pa (4.8 torr). Horizontal field width $= 350\,\mu m$. Reproduced from Rizzieri *et al.* (2003b), copyright American Institute of Physics

composite foams (Stokes and Donald, 2000), onion epidermal cells (Donald *et al.*, 2003) and geomaterials such as chalk (Sorgi and De Gennaro, 2007). Figure 6.12 is an example of an experiment on a hydrated specimen, and shows intercellular failure of onion epidermal cells.

6.3.2 Low-Temperature Experiments

In Chapter 3, Section 3.4, the properties of water were introduced and discussed in some detail. An extension to this is to work with small pressures of water vapour at sub-zero temperatures, taking experiments into the frozen-hydrated regime and opening up the possibility to observe phenomena such as changes of state (i.e. solid, liquid, gas) and phase transitions (e.g. transitions that have a sub-zero glass transition temperature T_g) as well as processes such as *in situ* freeze-drying. As described in Chapter 3, specimen equilibria can be determined from the phase diagram for water, and Figure 6.13 shows a plot of saturated water vapour pressure extended down to $-20\,°C$. At $0\,°C$, the three phases of water (vapour, liquid and solid) co-exist, at the triple point.

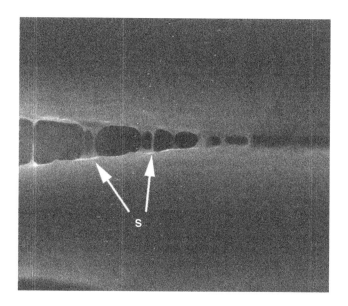

Figure 6.12 Real-time dynamic image showing intercellular failure of fully hydrated onion epidermis. There are two cells (top and bottom). Arrows mark the stringy features that participate in the adhesion of the cells and ultimately their debonding. Reproduced from Donald *et al.* (2003), copyright Oxford University Press

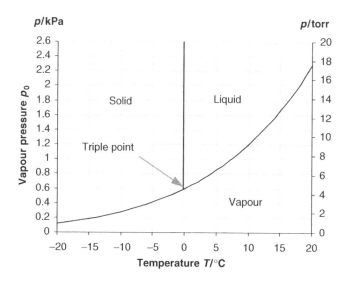

Figure 6.13 Extended phase diagram of water showing all three phases, liquid, solid and gas, as a function of temperature and pressure

Despite the excellent properties of water vapour as an imaging gas, an alternative gas is needed at temperatures lower than about $-60\,^\circ$C, since the equilibrium vapour pressure of water becomes vanishingly small and it is not possible to attain the very small pressures needed (below, say, 10 Pa) if we are to avoid precipitation of water as ice. Furthermore, such low pressures do not generally provide a high enough concentration of gas molecules for imaging and charge control.

By using different imaging gases, lower temperatures may be employed in cryo-VP-ESEM, typically up to $T = -150\,^\circ$C, for which sublimation of frozen-hydrated phases is negligible.[2] Studies have shown that good results can be obtained using nitrogen gas (Meredith *et al.*, 1996; Fletcher, 1997; Tricart *et al.*, 1997; Stokes *et al.*, 2004b). Nitrogen has a significant vapour pressure at low temperatures and so remains in the gaseous state. Nitrous oxide has also been demonstrated, and carbon dioxide can be used over a more limited temperature range.

Figure 6.14 shows the complex microstructure of frozen-hydrated ice cream, imaged using nitrogen as the chamber gas, during a study to

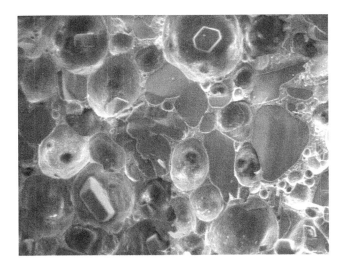

Figure 6.14   Secondary electron images of ice cream obtained at a temperature of $-95\,^\circ$C and imaged using nitrogen gas, pressure $p = 93$ Pa (0.7 torr). Primary beam energy $E_0 = 5$ keV. Horizontal field width $= 200\,\mu$m. Reproduced with permission from Stokes *et al.* (2004b). Copyright Blackwell Publishing

---

[2] Note that there is a high risk of the sublimation of ice at temperatures above about $T = -90\,^\circ$C. This means that there is currently a region of the phase diagram between $-90\,^\circ$C and $-60\,^\circ$C where conditions for thermodynamic stability of ice are not met in the VP-ESEM. However, short-term kinetic factors mean that this gap can be bridged if the conditions are changed very rapidly on passing through this region.

establish the conditions needed to determine stable imaging and control over thermodynamics (Stokes *et al.*, 2004b) using a specially adapted cryo-transfer system. This work has recently been further explored (Waller, 2007; Waller *et al*, 2008) and has successfully demonstrated the *in situ* freezing of solutions containing either pure water or anti-freeze proteins that affect ice crystal nucleation and growth, allowing both the liquid and solid phases to be visualised whilst using water vapour and/or nitrogen for imaging.

When working in the temperature range $-60\,°C$ and above, water vapour can and should be added to stabilise the ice in a frozen-hydrated specimen. But, as commented previously, the imaging gas pressure may be too low to enable sufficient signal amplification for image formation. Mixing another gas in with water vapour is a good way to increase the total pressure and, hence, the signal. One method for achieving this is described by Fletcher (1997), and involves apparatus to pre-mix different gases before addition to the specimen chamber. An alternative approach, crude but effective, is to have a base level of an auxiliary gas such as nitrogen and a starting temperature below the point where water vapour needs to be added (i.e. $T < -60\,°C$). Then, using the microscope's pressure readout to monitor the situation, raise the temperature to the desired level and carefully add the appropriate partial pressure of water vapour. If the temperature is to be raised further, then water vapour can be added in incremental steps, in line with the vapour pressure needed for specimen stability. Note that this only works in the direction of increasing temperature, since decreasing the temperature would require the removal of mixed gas, leaving uncertainty in the ratio of gases present.

This approach can also be used to grow ice crystals *in situ*, by deliberately raising the water vapour pressure to precipitate ice. Figure 6.16 shows an example, where the baseline imaging gas is nitrogen and a little water vapour has been added at a pressure above its saturated vapour pressure, causing precipitation of ice onto the cold surface, while nitrogen remains in the gas phase.

Experiments similar to that shown in Figure 6.15 have been utilised by Zimmermann *et al.* (2007) to assess the ice-nucleating properties of airborne particles. Water vapour was used in isolation as the imaging gas in that work, since the temperatures were in the range $-23\,°C$ to $-3\,°C$ where water vapour pressures are high enough that no additional gas is needed for imaging purposes.

Frozen fluid inclusions in various sedimentary minerals have been microanalysed using a cryostage in the VP-ESEM (Timofeeff *et al.*,

Figure 6.15 The onset of *in situ* crystallisation of ice on the surface of albite particles. Primary electron beam energy $E_0 = 15\,\text{keV}$. Water vapour pressure $p = 100$ Pa (0.75 torr). Temperature $T = -15\,°\text{C}$. Horizontal field width = 600 μm. Image courtesy of Martin Ebert, Technical University Darmstadt

2000). These inclusions are the remnants of fluid from which the crystal grew and give important clues about the chemical composition and temperature of water in the surface and subsurface environment that prevailed during the formation of the mineral.

6.3.3 High-Temperature Experiments

Another interesting application of VP-ESEM is to study the effect of high temperatures on microstructure and specimen chemistry, or to create nanomaterials *in situ*. This generally requires the use of specialised heating stages, some of which are capable of maintaining a temperature of up to 1500 °C. An early example of *in situ* heating in the VP-ESEM is that of Singler *et al.* (1993), who observed the behaviour of solder alloys as part of a correlative study on the spreading of solder on microelectronic circuits, using air, nitrogen and water vapour as imaging gases.

Alternatively, an electric current can be applied directly to the specimen so that it becomes resistively heated. For example, Wako *et al.* (2007)

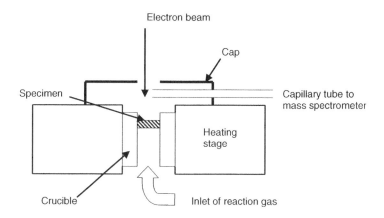

Figure 6.16 Simplified schematic diagram to show the delivery of a reactive gas to a heated specimen in the VP-ESEM. Diagram courtesy of Charlotte Appel, Haldor Topsøe

used ethanol vapour to grow single-walled carbon nanotubes on a silica substrate that had been resistively heated to 780 °C. A cobalt catalyst was distributed on the substrate surface and the authors were able to discern incubation, extension and termination growth stages of the nanotubes.

Several workers have used modified versions of heating stages to perform specific experiments. For example, a reactive gas can be fed to the heated specimen (Figure 6.16) to trigger a chemical reaction (Appel *et al.*, 2002; Klemenso *et al.*, 2006). Further, a mass spectrometer can form part of the system, so that reaction products can be detected and analysed. Another modification involves a probe tip attached to a micromanipulator, incorporated in order to agitate the specimen during sintering (see Smith *et al.*, 2006; Samara-Ratna *et al.*, 2007).

The arrangement shown schematically in Figure 6.16 has been used to observe the effects of *in situ* reduction and oxidation (redox) reactions of iron (Appel *et al.*, 2002). Oxidation was achieved by using water vapour as both the imaging gas and the reactive component of the system, at a specimen temperature $T = 300\,°C$. For the reduction reaction, a 50:50 mixture of hydrogen and carbon monoxide was used as the reactive component while imaging was carried out in a 60:40 helium/hydrogen mixture. In this case, the heating stage was held at a temperature $T = 500\,°C$. The results are shown in Figure 6.17.

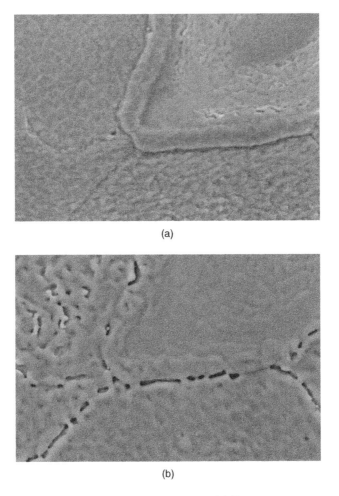

(a)

(b)

Figure 6.17 Images of the surface of iron obtained following an *in situ* experiment using the heating stage shown in Figure 6.16. The effect of oxidation in water vapour is shown in (a) while the effect of reduction in hydrogen/carbon monoxide is shown in (b). In (a) the formation of an oxide layer is found to depend on the underlying grain structure, with the layer being thicker at the grain boundary. Horizontal field width = 45 μm. Reproduced with permission from Appel *et al.* (2002)

Similarly, the modified gas flow stage in Figure 6.16 was used to study the redox process for nickel-yttria-stabilised zirconia, used as the anode for solid oxide fuel cells (Klemenso *et al.*, 2006). Oxidation was performed using air as both the imaging and reactive gas, while reduction involved hydrogen as the reactive gas, with argon acting as a carrier gas ($[H_2] = 9\%$ by volume). Redox cycling was carried out at temperatures between $T = 20\,°C$ and $T = 850\,°C$.

6.3.4 Condensation and Evaporation of Water

If the imaging gas is water vapour, then we have the ability to condense or evaporate water from the specimen, and perform a variety of *in situ* experiments. With reference to Figure 6.13, it is clear that if the pressure in the chamber is increased or the temperature of the sample is reduced, water will begin to condense as conditions shift to the liquid regime of the diagram. Likewise, lowering the chamber pressure or heating the sample will cause water to evaporate, as conditions shift to the gaseous or vapour regime of the diagram.

6.3.4.1 Wetting Experiments

Several workers have shown that useful information can be obtained by condensing water onto a substrate in order to assess the hydrophilic or hydrophobic tendencies of the material (for example: Cameron, 1994; Jenkins and Donald, 1997; Liukkonen, 1997; Jenkins and Donald, 1999; 2000; Stelmashenko *et al.*, 2001; Rossi *et al.*, 2004; Brugnara *et al.*, 2006). Specifically, by measuring the angle θ made between a droplet of water and the surface (the contact angle), it is possible to quantify the surface energy directly. Figure 6.18 shows the basic principle of contact angle measurement.

Contact angle measurement is normally done using a light optical system. In the VP-ESEM, where the depth of field is higher than for the light microscope and different mechanisms are available for contrast, observation of the contact point is perhaps more straightforward, provided that the specimen surface and water droplet are at a convenient viewing angle. A useful way to observe the contact angle on a bulk

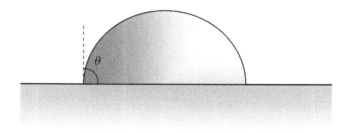

Figure 6.18 Schematic diagram depicting the measurement of a contact angle for a solid–water–air system on a flat surface

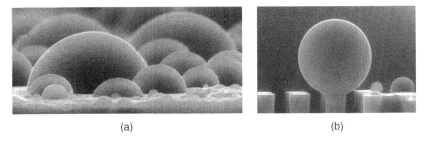

(a)   (b)

Figure 6.19 Viewing water droplets condensed *in situ* along the side of a bulk sub-strate (polystyrene and silicon, respectively). Horizontal field width = 70 µm in both cases. Image in (a) Reproduced from Stelmashenko *et al.* (2001), copyright Wiley-Blackwell. Image in (b) courtesy of M Leboeuf, University of Neuchatel

material is to position one edge of the substrate under the electron beam, to give a view along the side of the specimen and hence a side view of the water droplets. A couple of examples are shown in Figure 6.19.

Note that, provided that the droplet radius R does not exceed a certain value, effects due to the force of gravity F_g can be safely neglected since the surface tension force F_σ will dominate. In the case of water, surface tension is rather high, so it turns out that the droplet radius can be as much as about 850 µm before we run into problems with gravity. This can be deduced as follows:

Force F_σ due to surface tension σ:

$$F_\sigma = 2\sigma/R \qquad (6.1)$$

Force F_g due to gravity g:

$$F_g = 2R\rho g \qquad (6.2)$$

where ρ is the density of the material and $g = 9.81$ m/s^2.

Setting $F_\sigma = F_g$ so that the two forces are in balance gives:

$$2\sigma/R = 2R\rho g \qquad (6.3)$$

Finally, rearranging in terms of R:

$$R = (\sigma/\rho g)^{1/2} \qquad (6.4)$$

For water, $\sigma = 7 \times 10^{-3}$ N/m, $\rho = 10^3$ kg/m^3, hence the radius at which gravity begins to have any significant effect on water droplets is $R = 0.845$ mm or $\sim$850 µm. The high surface tension of water can also make it difficult to remove condensed water from small curved spaces such as pores.

Specimens are usually viewed from above, in plan view, and so it is not possible to see the contact point directly. However, Stelmashenko *et al.* (2001) and Brugnara *et al.* (2006) have formulated indirect methods for interpreting the secondary electron signal intensity across water droplets and then relating this to the contact angle. Liukkonen (1997) is one of several workers to use contact angle measurements in the VP-ESEM to study water transport in paper, while Lauri *et al.* (2006) correlate theoretical models of wetting with experimental observations on newspaper, Teflon and cellulose film.

Other work utilising *in situ* condensation of liquid water is that of White *et al.* (2006) for studying the reactions responsible for discoloration of paint pigments and Ma *et al.* (2006) for observing morphological changes of DL-alanine mesostructures at different humidities. Several types of material have been investigated by Rodriguez-Navarro *et al.* (2000) and Doehne (2006), correlating VP-ESEM observations of the hydration of stone, along with the deliquescence of salt crystals, which have important implications for the conservation of culturally important sites and in the preservation of art. Calcium carbonate dust particles have been similarly studied, and analysed using X-ray microanalysis. Laskin *et al.* (2005), Ebert *et al.* (2002) and Inerle-Hof *et al.* (2007) have observed and microanalysed the hygroscopic behaviour of aerosol particles, which are of great importance in connection with atmospheric chemistry.

Wirth *et al.* (2008) have investigated the adhesive properties and effects of wetting on vertically aligned carbon nanotube arrays by condensing water onto the surface of the array. This is found to lead to a loss of hydrophobicity, which can be related to the compaction of the nanotube arrays by the water.

Rossi *et al.* (2004) have used *in situ* VP-ESEM to visualise water transport in carbon nanotubes, utilising electron beam penetration to 'see' through the walls of the nanotubes. Their results are comparable to those obtained on sealed nanotubes in the TEM. By observing the fluid dynamics of water inside the nanotubes and looking at the angles of liquid menisci in contact with the interiors of the nanotubes, they were able to conclude that, unusually, the nanotubes they had produced were hydrophilic. The results of this work are shown in Figure 6.20.

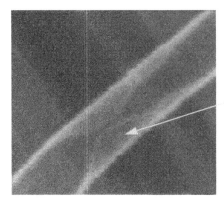

Water inside
carbon nanotube

Figure 6.20   Observation of the movement of water in carbon nanotubes in the VP-ESEM. Horizontal field width = 1 μm. Reproduced with permission from Rossi *et al.* (2004). Copyright American Chemical Society

6.3.4.2 *Controlled Evaporation*

Evaporation of solvent *in situ* is a very good way to observe the behaviour of, for example, colloidal systems. Colloids[3] are very small particles that have dimensions of a few microns or less, dispersed in another medium, and exhibit very interesting properties.

Colloidal systems tend to be governed by short-range interactions: if particles are brought close enough together, they can form specific structures, either by staying a little way apart or by aggregating in some way. One example is that of a well-ordered structure known as a colloidal crystal (the term still applies even if the particles are in a liquid). Aside from existing all around us in everyday life, colloids can have an important role to play in many processes. For example, they control the drying of paint and the properties of foods and can be used as templates in the fabrication of nanophotonic devices.

Figure 6.21 schematically shows the physical principles involved in the formation of a colloidal crystal. Once within physical range of the secondary potential energy minimum, particles are attracted to each other. But the primary maximum prevents the colloidal particles from actually sticking together, unless a large amount of energy is supplied to the system or the potential energy is lowered (reducing the height of the primary maximum) by the addition of an electrolyte.

---

[3] Some examples of colloids: opal, paint, toothpaste, milk, jelly, cheese, mayonnaise, fog and smoke.

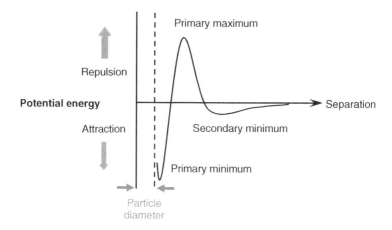

Figure 6.21   Schematic plot of potential energy with particle separation to illustrate long-range repulsion (energy maxima) and short-range attraction (energy minima) in a colloidal system

An early example of using VP-ESEM to observe this phenomenon at the nanoscale is given by He and Donald (1996). Figure 6.22 is the result of an experiment in which the concentration of the aqueous phase has been carefully reduced, causing the colloidal particles to be close enough to feel the short-range attractive potential described above.

A number of similar studies have been carried out to observe the formation of films, often of great importance to the paint manufacturing industry (Keddie *et al.*, 1995; He and Donald, 1996; Keddie *et al.*, 1996a, 1996b; Meredith and Donald, 1996; Stelmashenko and Donald, 1998; Donald *et al.*, 2000; Dragnevski and Donald, 2008).

6.3.4.3 *Wet–Dry Cycling*

Experiments involving control over chamber water vapour pressure to cycle between the wet and dry states have been conducted by Bache and Donald (1998), who were able to observe the development of a network structure in the wheat protein gluten during careful removal of water. They also noticed that the history of the specimen during and after pumpdown could influence the resultant microstructure.

Also, Jenkins and Donald (1997) took cross-sections of textile fibres, with a view to measuring swelling and shrinkage in the hydrated and dehydrated states, respectively. A specialised mounting technique was used to observe the transverse dimensions of the fibres, allowing good

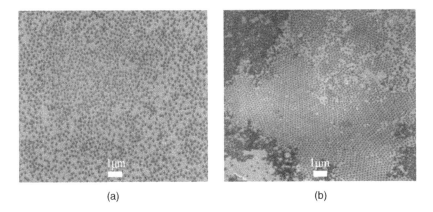

(a)   (b)

Figure 6.22 Latex nanoparticles ($\sim$100 nm radius) undergoing *in situ* colloidal crystallisation. In (a) the particles distribute themselves at random: there is no particular ordering of particles. In (b), after removal of solvent (water) by controlled evaporation, the particles are brought closer together. Short-range forces cause the particles to adopt a crystalline habit. Horizontal field width = 11 μm. Reproduced from He and Donald (1996), copyright American Chemical Society

thermal contact between the Peltier-cooled specimen holder and the fibres. In this work, conditions are also described for controlling radiation damage in the presence of water.

6.3.5 Processes using Electron Beam Gas Chemistry in the VP-ESEM

6.3.5.1 *Electron Beam Lithography*

As we have already seen, the action of the electron beam in a gaseous environment opens up some interesting possibilities. The primary electron beam can be used to our advantage to initiate specific chemical reactions in the near-surface bulk of materials, giving rise to processes such as cross-linking, akin to the mechanisms outlined in the section on radiation damage in Chapter 2, Section 2.5.1. This can result in the formation of micro- and nanoscale features, and is the basis of electron beam lithography, which is used to write structures into materials such as polymers to make resistive structures (usually carried out in a high-vacuum environment).

A key factor in lithography is control over critical dimensions, just as we discussed in the section on nanometrology (Section 6.2.3). Of course, for polymeric substrates, charging causes drifting of the primary electron beam in the high-vacuum SEM, which ultimately increases the thickness of electron beam-written lines. Recently, Myers and Dravid (2006) have shown that VP-ESEM can be used to control these artefacts and that, in the absence of charging effects, highly accurate electron beam nanolithographic features can be created with smaller line widths than for the high-vacuum case. The influence of the primary electron beam skirt was also investigated and the only effect found was that the required electron dose needed to be increased relative to that used under high-vacuum conditions. This further helps to support the idea that the skirt is not a significant factor in determining resolution.

Toth *et al.* (2007b) also explore the potential of VP-ESEM to create finer structures than currently possible with conventional lithography techniques by post-etching beam-deposited structures. The process of deposition is discussed in the next section.

6.3.5.2 In Situ *Nanofabrication*

The interactions of electrons with certain gases can lead to the deliberate deposition of metals and other materials, giving rise to structures that are said to be the result of *in situ* chemical vapour deposition (CVD). The fabrication of gold nanotips has been shown by Molhave *et al.* (2003), via decomposition of the precursor gas dimethylacetylacetonate gold (III), $[Au(CH_3)_2(C_5H_7O_2)]$. Experiments were carried out using either nitrogen or water vapour as the imaging gas in the VP-ESEM, and the nanostructures of the resultant deposits were found to vary in composition depending on the choice of imaging gas. For a nitrogen environment with $p = 133$ Pa (1 torr), gold nanocrystals were formed in an amorphous carbon matrix, while for a water vapour environment with $p = 120$ Pa (0.9 torr) the tips consisted of a dense gold core surrounded by a layer of material similar to that found for the previous result. This is shown in Figure 6.23. Madsen *et al.* (2003) have used gold deposition to demonstrate how this approach can be used to accurately solder carbon nanotubes onto microelectrodes, thereby creating an electrical connection.

As already discussed in Section 6.3.3, carbon nanotubes can be grown in an organic vapour environment in conjunction with a heating stage. Other examples of this type of work include the growth of diamond nanofilms (Niitsuma *et al.*, 2006) and carbon nanotubes/films (Niitsuma *et al.*, 2005; Niitsuma *et al.*, 2007). Toth *et al.* (2007a) have studied the competition between beam-induced etching and deposition in the

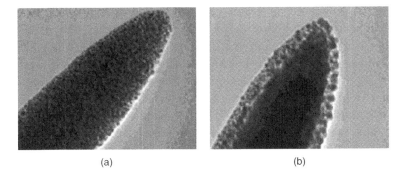

(a)   (b)

Figure 6.23 *In situ* nanofabrication in the VP-ESEM. (a) A TEM image showing a nanotip deposited from a gold precursor gas in the presence of nitrogen. Gold nanoparticles are dispersed in an amorphous carbon matrix. Meanwhile (b) was deposited in a water vapour environment, leading to a more dense gold core surrounded by a thin amorphous layer of material similar to that in (a). Horizontal field width = 160 nm. Reproduced from Molhave *et al.* (2003), copyright American Chemical Society

VP-ESEM using the reactive gas xenon difluoride to mediate these processes on lithographic masks and carbonaceous films.

The topics discussed in this section are relatively new to the VP-ESEM and should be of great interest in nanotechnology and nano-science as well as for the semiconductor industry. This is an area that clearly holds great promise for the future.

6.3.6 High-Pressure Experiments

As mentioned, there is still scope for reaching new milestones and finding new applications, even though VP-ESEM technology has reached a certain level of maturity. Another example is the goal of imaging and analysis at physiological temperatures and pressures, which remains to be demonstrated. However, progress has been made, showing that true secondary electron imaging is possible at high pressures (Toth and Baker, 2004; Stokes *et al.*, 2004a). An example is shown in Figure 6.24, where an evaporated tin ball standard has been imaged with water vapour pressure $p = 2.66$ kPa (20 torr). Using a short gas path length, as discussed in Chapter 4, a relatively intense local electric field helps to extract the secondary electrons to the anode. The physical principles are described in Toth *et al.* (2007c). The real test is to image a soft, organic material in this way, and a preliminary example (live cyanobacteria) is shown in Figure 6.25 (see also Stokes, 2006).

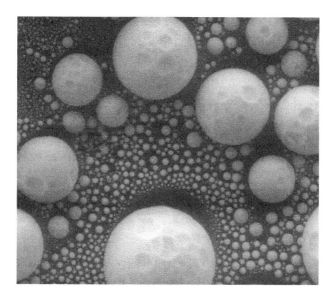

Figure 6.24 Imaging of tin balls at a water vapour pressure $p = 2.66\,\text{kPa}$ (20 torr). Primary beam energy $E_0 = 20\,\text{keV}$. Horizontal field width = $55\,\mu\text{m}$. Note the appearance of surface detail, indicating that the signal contains a significant secondary electron component. Reproduced with permission from Stokes (2006). Copyright Royal Microscopical Society

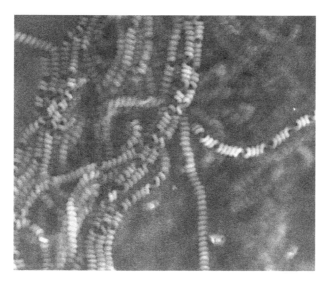

Figure 6.25 A biological specimen (cyanobacteria *spirlunina*) imaged at room temperature, water vapour pressure $p = 14.5$ torr (65 % RH). Horizontal field width = $50\,\mu\text{m}$. Sample courtesy of David Adams, University of Leeds

6.4 OTHER APPLICATIONS

6.4.1 Introduction

In this final section, a range of applications is surveyed, to further demonstrate the current breadth and depth of the work being carried out using VP-ESEM. This is a brief summary – it is not exhaustive. The aim is simply to highlight the different specimen types and application areas that have so far been examined, along with some representative images, to give the reader an idea of the possibilities and a starting point in the literature.

6.4.2 Biological Specimens

Imaging of biological specimens is another challenging area in electron microscopy, with specimens being electrically insulating, radiation- and vacuum-sensitive and often involving internal aqueous media and external secretions. Of course, it is appreciated that many excellent techniques have been developed for the purposes of studying biological specimens using high-vacuum SEM. Nonetheless, there are a growing number of examples of VP-ESEM in the life sciences, where minimal specimen preparation and hence the 'natural' state is a key consideration. Undoubtedly, though, more work is needed in understanding the implications of imaging biological materials without a conductive coating if we are to move towards experiments involving living systems under physiological conditions, for example.

In Chapter 3, Section 3.4 it was stated that the equilibrium conditions (water vapour pressure and temperature) for a specimen consisting of an aqueous medium correspond to somewhat less than 100 % humidity, relative to pure water, and that the use of too high a vapour pressure could result in unwanted condensation of water onto the specimen surface. This is borne out by the observations of several groups. Tai and Tang (2001), for example, noted that for biological tissues, water vapour pressures above about 90 % RH lead to the formation of water droplets on the surface. For other instances demonstrating the successful handling of cells and tissues by observing these principles see, for example, Stokes *et al.* (2002); MacKinlay *et al.* (2004); Bergmans *et al.* (2005) and Muscariello *et al.* (2005). Figure 6.26 shows chemically fixed but uncoated, semi-hydrated osteoblasts (bone cells) on glass.

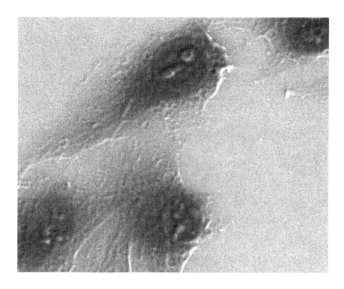

Figure 6.26   Secondary electron image of uncoated cells in the VP-ESEM, imaged using water vapour with $p \sim 385$ Pa (2.9 torr), temperature $T = 3\,°C$ and primary beam energy $E_0 = 10\,keV$. Horizontal field width $\sim 110\,\mu m$. Reproduced with permission from Stokes *et al.* (2002). Copyright John Wiley and Sons, Inc

Muscariello *et al.* (2005) and Bergmans *et al.* (2005) have given excellent reviews on the use of VP-ESEM for biological materials, along with some very nice results. For a complementary approach to imaging cells utilising confocal laser scanning microscopy, high-vacuum SEM and VP-ESEM, making the most of the various advantages of each, the reader is referred to MacKinlay *et al.* (2004). A review on the application of VP-ESEM to the imaging of soft materials can be found in Donald *et al.* (2000).

Other biological examples in the VP-ESEM include studies of rat chondrocytes and epithelial cells (Cohen *et al.*, 2003) and rat intestinal mucosa (Habold *et al.*, 2003), as well as the attachment and growth of endothelial cells on hollow-fibre capillaries (Neuhaus *et al.*, 2006). Gedrange *et al.* (2005) performed X-ray microanalysis on muscle tissue in a VP-ESEM using tissue taken from animals that had been treated with Botox. The experiments were used to determine ionic concentrations in order to deduce whether these can be used as indicators of muscle activity.

Misirli *et al.* (2007) give a comparative study between imaging yeast cells with and without temperature control to maintain the hydrated state

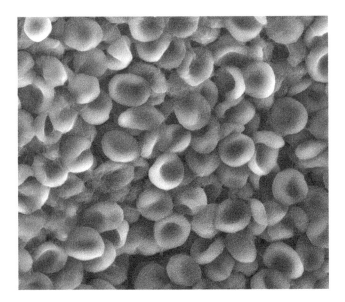

Figure 6.27 VP-ESEM image of hydrated clotted erythrocytes (red blood cells). Temperature $T = 6\,^\circ\text{C}$, water pressure $p \sim 470$ Pa (3.5 torr). Reproduced from Iliescu *et al.* (2008), copyright John Wiley & Sons, Inc

in the VP-ESEM, and Iliescu *et al.* (2008) report on chitosan-glycerol phosphate/blood implants being developed for cartilage repair via a clotting mechanism, which was imaged and analysed using EDS in the VP-ESEM. An example of this work is shown in Figure 6.27. The calcification of porcine prosthetic heart valves has been reported by Delogne *et al.* (2007).

The application of VP-ESEM to bacteria, fungi, moss and lichen can be studied in papers by Callow *et al.* (2003); Castillo *et al.* (2005); Weimer *et al.* (2006) and Basile *et al.* (2008). Specifically, Weimer *et al.* (2006) used VP-ESEM to understand the mechanisms of bacterial adhesion to cellulose, similar to that shown in Figure 6.28, while Basile *et al.* (2008) used VP-ESEM combined with X-ray microanalysis to study moss and lichen. Bamboo root and leaf specimens have been observed by Lux *et al.* (2003), and Micic *et al.* (2000) have followed the supramolecular organisation of enzymatically polymerised lignin on graphite, mica and glass substrates. Biofilms are known to interfere with mineral barriers designed to reduce pollution and are notoriously difficult to analyse by conventional SEM and TEM. However, VP-ESEM has been successfully used to understand microbial interactions that lead to the formation

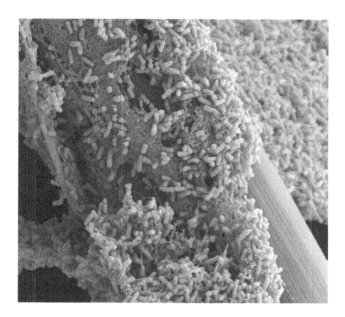

Figure 6.28 VP-ESEM image of bacterial biofilm of *Pseudomonas aeruginosa* Primary beam energy $E_0 = 5\,$keV. Horizontal field width $= 32\,\mu$m. Image courtesy of Paul Gunning, Smith and Nephew Technology Services Group

of barrier-clogging biofilms (Darkin *et al.*, 2001). Likewise, sludge flocs have been studied by Holbrook *et al.* (2006).

It is notable that several of the authors mentioned in this section (Callow *et al.*, Bergmans *et al.* and Misirli *et al.*) point out that there is a significant gain in morphological information from imaging hydrated specimens, even if the structures are partially obscured by surrounding aqueous media, since the conditions are more representative of the specimen's real environment and therefore give additional clues as to structure–property relations and local interactions.

There are very few papers on X-ray microanalysis of hydrated biological specimens, although an early example is that of Egertonwarburton *et al.* (1993), which discusses the difficulties of analysing material that has not been immobilised by drying or freezing, and hence introduces the risk of ion migration that may influence metal localisation in tissues. Surface roughness is also an issue, and can result in the absorption of X-ray signals affecting the collection of quantitative data. Gilpin and Sigee (1995) recognised the importance of working distance (gas path length) in X-ray microanalysis of hydrated biological specimens.

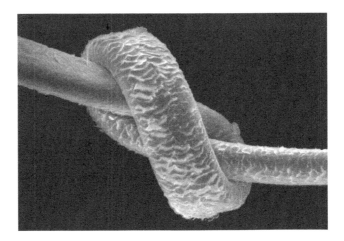

Figure 6.29   A knotted strand of human hair. Water vapour pressure $p = 173$ Pa
(1.3 torr) Horizontal field width $= 380\,\mu m$. Image courtesy of Juliette McGregor,
University of Cambridge

The VP-ESEM also finds use in forensic analysis, particularly because
of the capability for imaging uncoated bulk specimens such as the human
hair shown in Figure 6.29. Saville *et al.* (2007) developed a means to
correlate the teeth on a saw to the marks made on bone inflicted as
a result of dismemberment, and were able to determine which marks
corresponded to the pushing of the saw and which to pulling.

VP-ESEM helped in the reconstruction of the last days of an ancient
man whose frozen body was found in a North American glacier. He is
thought to have been there for around 550–600 years (Dickson *et al.*,
2004). Analysis of stomach contents and particles such as pollen on
clothing indicated that the man had probably lived near the coast and
was travelling inland before his death.

6.4.3 Liquids and Soft Materials

As we have already seen, the imaging capabilities of the VP-ESEM
include the ability to observe complex fluids. For example, oil/water
emulsions can be stabilised and imaged provided that the minimum
conditions (temperature and pressure) for metastability of the water are
met, as described in Chapter 3, Section 3.4. But water vapour only
stabilises water, so what about the stability of other liquid phases in
the absence of a suitable vapour? Well, that depends on their vapour
pressures. For example, oil phases typically have a considerably lower

vapour pressure than water and so there is no *appreciable* evaporation of the liquid oil phase (Stokes *et al.*, 1998).

Other types of liquid were mentioned in Sections 6.2.5 and 6.3.4.2. These included wetSTEM imaging of emulsions and suspensions and several *in situ* studies on the mechanisms of film formation during the drying of fluid-based materials such as paints.

An early example of imaging soft materials in the VP-ESEM is that of Tricart *et al.* (1997), for imaging oil-based mud. Meanwhile, the confinement of liquid crystals in carbon nanotubes has been studied in VP-ESEM by Shah *et al.* (2006). This is a subject of interest in the field of optoelectronics. Similar to the work of Rossi *et al* mentioned in Section 6.3.4.1, this involved nanoscale observation of fluid menisci inside carbon nanotubes.

Polymer blends have been studied in the VP-ESEM by Bache *et al.* (2000) and Williams *et al.* (2005), as have biopolymer gels (Bache *et al.*, 1997; Plucknett *et al.*, 2001). The morphology of a photo-switchable polyethylene glycol-based hydrogel was reported by Micic *et al.* (2003) and electrospun polyacrylonitrile fibres for use as artificial muscle were detailed by Samatham *et al.* (2006). The latter showed the effect of pH on the diameter of fibres, which were studied in water, hydrochloric acid and sodium chloride solution. This was combined with data from atomic force microscopy. Royall *et al.* (2001) carried out a complementary study using confocal laser scanning microscopy and VP-ESEM for silica-based polymer lacquers in addition to complex fluids.

Crack propagation and failure of elastomers has been studied by Hainsworth (2007), along with the properties of organic layers that form automotive paint coatings (Hainsworth and Kilgallon, 2008), and various paint pigments have been studied by Cavallo (2006), Doehne (2006), White *et al.* (2006) and Chiari *et al.* (2008). An example of the latter work is shown in Figure 6.30. These delicate fibres are the product of nanotechnology dating back to the ancient Mayan civilisation.

VP-ESEM is also well suited for applications involving the microstructure of pharmaceutical materials and foods. For example, pharmaceutical applications include the swelling and dissolution of drug-carrier particles of different formulations (Roberts *et al.*, 1997). Carlton (2006) discusses the use of VP-ESEM imaging and X-ray microanalysis of pharmaceuticals for morphological studies and mapping of the distributions of components as well as for determining and identifying any contaminants present. Several food systems have been covered in this chapter and have included vegetable tissues (carrot, onion), biopolymers

Figure 6.30 VP-ESEM image of nanofibres of the pigment Maya Blue. Horizontal field width = 1.6 μm. Reproduced from Chiari *et al.* (2008), copyright the J. Paul Getty Trust. All rights reserved

(gelatin, dextran, maltodextrin, breadcrumb) and emulsions (mayonnaise, ice cream). Observations in the VP-ESEM can be linked to the materials science of foods, and this has been reviewed by Donald (2004).

6.4.4 Hard/Soft Composites and Hard Materials

Moving towards some harder materials, Dusevich and Eick (2002) discuss the risk of shrinkage from the fracture preparation of nondehydrated demineralised dentin specimens for the VP-ESEM, where they recommend fixing in order to overcome the problem. Also on the subject of dentistry, Franz *et al.* (2006) studied heterogeneous glass/polymer composites for use as synthetic tooth materials. The interface between such materials and the tooth is crucial in determining good integration and adhesion. This was investigated for tooth–composite explants, along with composite–composite specimens. Figure 6.31 shows an example of the interface between two composites.

A similarly important system for biomedical applications is the calcium phosphate bone analogue hydroxyapatite (see, for example, Botelho *et al.*, 2006), and an example is shown in Figure 6.32. Plasma sprayed hydroxyapatite coatings have been characterised using both VP-ESEM and cathodoluminescence (Gross and Phillips, 1998).

Figure 6.31   VP-ESEM study of the interface between two glass–polymer composites showing differences in morphology. Arrow marks the interface. Field of view = 130 μm. Reproduced from Franz *et al.* (2006), copyright Springer

Figure 6.32   VP-ESEM image of the surface of a calcium phosphate ceramic material (hydroxyapatite), typically used as a bioactive bone analogue for surgical implants. Small grains can be seen in the image, which was obtained using water vapour with $p \sim 330$ Pa (2.5 torr) and primary beam energy $E_0 = 20$ keV. Horizontal field width = 25 μm

Meanwhile, the alkali–silica reaction responsible for degradation of concrete has been investigated using X-ray microanalysis (Verstraete *et al.*, 2004; Khouchaf and Verstraete, 2007). Further, Khouchaf and Boinski (2007) provide a comparison on using helium and water vapour for X-ray microanalysis of silica, and Wei *et al.* (2003) demonstrate this approach for ceramic fibres. Similarly, cement-based materials have been studied by Hall *et al.* (1995).

In addition to some of the techniques such as X-ray microanalysis and cathodoluminescence mentioned in this chapter, it has also been shown that electron backscatter diffraction (EBSD) is possible in the gaseous environment of the VP-ESEM (see, for example, Thaveeprungsriporn *et al.*, 1994). And finally, it is noted that a TEM-compatible transfer system for the VP-ESEM has been demonstrated by Kaegi and Holzer (2003), for handling the transfer of particulate specimens.

REFERENCES

Appel, C.C., Rasmussen, A.-M., Ullmann, S. and Hansen, P.L. (2002). *Experiments with a modified heating stage on an environmental scanning electron microscope.* ICEM-15, Durban, South Africa.

Bache, I.C., Anderson, V.J., Jones, R.A.L. and Donald, A.M. (1997). *The Observation of Hierarchical Structures in Biopolymer Phase Separation: Novel ESEM Contrast Mechanisms.* Microscopy & Microanalysis '97, Cleveland, OH.

Bache, I.C. and Donald, A.M. (1998). The Structure of the Gluten Network in Dough: A Study Using Environmental Scanning Electron Microscopy, *J. Cereal Sci.,* **28**, 127–133.

Bache, I.C., Thomas, D.S. and Donald, A.M. (2000). *Analysis of Conjugated Polymer Multilayer Structures in the Environmental SEM.* Institute of Physics Conference Series 165 (Symposium 7).

Baroni, T.C., Griffin, B.J., Browne, J.R. and Lincoln, F.J. (2000). Correlation Between Charge Contrast Imaging and the Distribution of Some Trace Level Impurities in Gibbsite. *Microsc. Microanal.,* **6**, 49–58.

Baroni, T.C., Griffin, B.J., Cornell, J.B., Roach, G.I.D. and Lincoln, F.J. (2002). Three-dimensional reconstruction of microstructures in gibbsite using charge contrast images. *Scanning,* **24**(1), 18–24.

Basile, A., Sorbo, S., Aprile, G., Conte, B. and Cobianchi, R.C. (2008). Comparison of the heavy metal bioaccumulation capacity of an epiphytic moss and an epiphytic lichen. *Environ. Poll.,* **151**(2), 401–407.

Bergmans, L., Moisiadis, P., Van Meerbeek, B., Quirynen, M. and Lambrechts, P. (2005). Microscopic observation of bacteria: review highlighting the use of environmental SEM. *Int. Endodont. J.* **38**(11), 775–788.

Bogner, A., Jouneau, P.H., Thollet, G., Basset, D. and Gauthier, C. (2007). A history of scanning electron microscopy developments: Towards "wet-STEM" imaging. *Micron,* **38**(4), 390–401.

Bogner, A., Thollet, G., Basset, D., Jouneau, P.H. and Gauthier, C. (2005). Wet STEM: A new development in environmental SEM for imaging nano-objects included in a liquid phase. *Ultramicrosc.,* **104**(3–4), 290–301.

Botelho, C.M., Brooks, R.A., Best, S.M., Lopes, M.A., Santos, J.D., Rushton, N. and Bonfield, W. (2006). Human osteoblast response to silicon-substituted hydroxyapatite. *J. Biomed. Mat. Res. A,* **79A**(3), 723–730.

Brugnara, M., Della Volpe, C., Siboni, S. and Zeni, D. (2006). Contact angle analysis on polymethylmethacrylate and commercial wax by using environmental scanning electron microscope. *Scanning*, 28, 267–273.

Callow, J.A., Osborne, M.P., Callow, M.E., Baker, F. and Donald, A.M. (2003). Use of environmental scanning electron microscopy to image the spore adhesive of the marine alga *Enteromorpha* in its natural hydrated state. *Colloids Surf. B-Biointerfaces*, 27(4), 315–321.

Cameron, R.E. (1994). Environmental Scanning Electron Microscopy in Polymer Science. *Trends Pol. Sci.*, 2(4), 116–119.

Carlton, R.A. (2006). Pharmaceutical applications of energy-dispersive spectroscopy in low-vacuum-environmental scanning electron microscopy. *Scanning*, 28(2), 108–109.

Castillo, U., Myers, S., Browne, L., Strobel, G., Hess, W.M., Hanks, J. and Reay, D. (2005). Scanning electron microscopy of some endophytic streptomycetes in Snakevine – *Kennedia nigricans*. *Scanning*, 27, 305–311.

Cavallo, G. (2006). The blue pigment used in Vallemaggia (Switzerland) in the half of 19(th) century by painters Vanoni and Pedrazzi. *Microchim. Acta*, 155(1–2), 121–124.

Chiari, G., Giustetto, R., Druzik, J., Doehne, E. and Ricchiardi, G. (2008). Pre-columbian nanotechnology: Reconciling the mysteries of the Maya blue pigment. *Appl. Phys. A – Mat. Sci. Proc.*, 90(1), 3–7.

Clausen, C. and Bilde-Sorensen, J. (1992). Observation of voltage contrast at grain boundaries in YSZ. *Micron Microsc. Acta*, 23(1/2), 157–158.

Clode, P.L. (2006). Charge contrast imaging of biomaterials in a variable pressure scanning electron microscope. *J. Struct. Biol.*, 155(3), 505–511.

Cohen, M., Klein, E., Geiger, B. and Addadi, L. (2003). Organization and adhesive properties of the hyaluronan pericellular coat of chondrocytes and epithelial cells. *Biophys. J.*, 85(3), 1996–2005.

Craven, J.P., Baker, F.S., Thiel, B.L. and Donald, A.M. (2002). Consequences of Positive Ions upon Imaging in Low vacuum SEM. *J. Microsc.*, 205(1), 96–105.

Cuthbert, S.J. and Buckman, J.O. (2005). Charge contrast imaging of fine-scale microstructure and compositional variation in garnet using the environmental scanning electron microscope. *Am. Mineral.*, 90(4), 701–707.

Darkin, M.G., Gilpin, C., Williams, J.B. and Sangha, C.M. (2001). Direct wet surface imaging of an anaerobic biofilm by environmental scanning electron microscopy: application to landfill clay liner barriers. *Scanning*, 23, 346–350.

Delogne, C., Lawford, P.V., Habesch, S.M. and Carolan, V.A. (2007). Characterization of the calcification of cardiac valve bioprostheses by environmental scanning electron microscopy and vibrational spectroscopy. *J. Microsc.–Oxford*, 228(1), 62–77.

Dickson, J.H., Richards, M.P., Hebda, R.J., Mudie, P.J., Beattie, O., Ramsay, S., Turner, N.J., Leighton, B.J., Webster, J.M., Hobischak, N.R., Anderson, G.S., Troffe, P.M. and Wigen, R.J. (2004). Kwaday Dan Ts'inchi, the first ancient body of a man from a North American glacier: Reconstructing his last days by intestinal and biomolecular analyses. *Holocene*, 14(4), 481–486.

Doehne, E. (1998). Charge Contrast: Some ESEM Observations of a New/Old Phenomenon. *Microsc. Microanal.*, 4(Suppl.2: Proceedings), 292–293.

Doehne, E. (2006). ESEM applications: From cultural heritage conservation to nano-behaviour. *Microchim. Acta*, **155**(1–2), 45–50.

Doehne, E. and Baken, E. (2006). An environmental STEM detector for ESEM: New applications for humidity control at high resolution. *Scanning*, **28**(2), 103–104.

Donald, A.M. (2004). Food for thought. *Nature Materials*, **3**, 579–581.

Donald, A.M., Baker, F.S., Smith, A.C. and Waldron, K.W. (2003). Fracture of plant tissues and walls as visualized by environmental scanning electron microscopy. *Ann. Botany*, **92**(1), 73–77.

Donald, A.M., He, C., Royall, C.P., Sferrazza, M., Stelmashenko, N.A. and Thiel, B.L. (2000). Applications of Environmental Scanning Electron Microscopy to Colloidal Aggregation and Film Formation. *Colloids Surf. A: Physicochem. Eng. Aspects*, **174**, 37–53.

Dragnevski, K.I. and Donald, A.M. (2008). An environmental scanning electron microscopy examination of the film formation mechanism of novel acrylic latex. *Colloids Surf. A: Physicochem. Eng. Aspects*, **317**(1–3), 551–556.

Dusevich, V.M. and Eick, J.D. (2002). Evaluation of demineralised dentin contraction by stereo measurements using environmental and conventional scanning electron microscopy. *Scanning*, **24**, 101–105.

Ebert, M., Inerle-Hof, M. and Weinbruch, S. (2002). Environmental scanning electron microscopy as a new technique to determine the hygroscopic behaviour of individual aerosol particles. *Atmos. Environ.*, **36**(39–40), 5909–5916.

Egertonwarburton, L.M., Griffin, B.J. and Kuo, J. (1993). Microanalytical Studies Of Metal Localization In Biological Tissues By Environmental SEM. *Microsc. Res. Techn.*, **25**(5–6), 406–411.

Fletcher, A.L. (1997). *Cryogenic Developments and Signal Amplification in Environmental Scanning Electron Microscopy*. Dept of Physics. University of Cambridge.

Franz, N., Ahlers, M.O., Abdullah, A. and Hohenberg, H. (2006). Material-specific contrast in the ESEM and its application in dentistry. *J. Mat. Sci.*, **41**(14), 4561–4567.

Gauvin, R., Robertson, K. and LeBerre, J.-F. (2003). Possibility of charge contrast imaging of polymeric materials. *Scanning*, **25**, 240–242.

Gedrange, T., Mai, R., Richter, G., Wolf, P., Lupp, A. and Harzer, W. (2005). X-ray microanalysis of elements in the masticatory muscle after paresis of the right masseter. *J. Dental Res.*, **84**(11), 1026–1030.

Gilpin, C. and Sigee, D.C. (1995). X-Ray-Microanalysis Of Wet Biological Specimens In The Environmental Scanning Electron-Microscope.1. Reduction Of Specimen Distance Under Different Atmospheric Conditions. *J. Microsc. – Oxford*, **179**, 22–28.

Goodhew, P.J., Humphreys, F.J. and Beanland, R. (2001). *Electron Microscopy and Analysis*, third edition, Taylor and Francis.

Griffin, B.J. (1997). A New Mechanism for the Imaging of Non-conductive Materials: An Application of Charge-Induced Contrast in the Environmental Scanning Electron Microscope (ESEM). *Microsc. Microanal.*, **3**(Supplement 2: Proceedings), 1197–1198.

Griffin, B.J. (2000). Charge Contrast Imaging of Material Growth Defects in Environmental Scanning Electron Microscopy – Linking Electron Emission and Cathodoluminescence. *Scanning*, **22**, 234–242.

Gross, K.A. and Phillips, M.R. (1998). Identification and Mapping of the Amorphous Phase in Plasma-sprayed Hydroxyapatite Coatings Using Scanning Cathodoluminescence Microscopy. *J. Mat. Sci.: Mat. Medicine*, **9**, 797–802.

Habold, C., Dunel-Erb, S., Chevalier, C., Laurent, P., Le Maho, Y. and Lignot, J.F. (2003). Observations of the intestinal mucosa using environmental scanning electron microscopy (ESEM); comparison with conventional scanning electron microscopy (CSEM). *Micron*, **34**(8), 373–379.

Hainsworth, S.V. (2007). An environmental scanning electron microscopy investigation of fatigue crack initiation and propagation in elastomers. *Polym. Testing*, **26**(1), 60–70.

Hainsworth, S.V. and Kilgallon, P.J. (2008). Temperature-variant scratch deformation response of automotive paint systems. *Prog. Organic Coatings*, **62**(1), 21–27.

Hall, C., Hoff, W.D., Taylor, S.C., Wilson, M.A., Yoon, B., Reinhardt, H.-W., Sosoro, M., Meredith, P. and Donald, A.M. (1995). Water anomaly in capillary liquid absorption by cement-based materials. *J. Mat. Sci. Lett.*, **14**, 1178–1181.

Harker, A.B., Howitt, D.G., Denatale, J.F. and Flintoff, J.F. (1994). Charge-Sensitive Secondary-Electron Imaging Of Diamond Microstructures. *Scanning*, **16**(2), 87–90.

He, C. and Donald, A.M. (1996). Morphology of Core-Shell Polymer Latices during Drying. *Langmuir*, **12**(26), 6250–6256.

He, C. and Donald, A.M. (1997). Morphology of a deformed rubber toughened poly(methyl methacrylate) film under tensile strain. *J. Mat. Sci.*, **32**, 5661–5667.

Holbrook, R.D., Wagner, M.S., Mahoney, C.M. and Wight, S.A. (2006). Investigating activated sludge flocs using microanalytical techniques: Demonstration of environmental scanning electron microscopy and time-of-flight secondary ion mass spectrometry for wastewater applications. *Water Env. Res.*, **78**(4), 381.

Horsewell, A. and Clausen, C. (1994). *Voltage contrast of ceramics in the environmental SEM*. ICEM 13, Paris, France.

Iliescu, M., Hoemann, C.D., Shive, M.S., Chenite, A. and Buschmann, M.D. (2008). Ultrastructure of hybrid chitosan–glycerol phosphate blood clots by environmental scanning electron microscopy. *Microsc. Res. Techn.*, **71**(3), 236–247.

Inerle-Hof, M., Weinbruch, S., Ebert, M. and Thomassen, Y. (2007). The hygroscopic behaviour of individual aerosol particles in nickel refineries as investigated by environmental scanning electron microscopy. *J. Environ. Monitor.*, **9**(4), 301–306.

Jenkins, L.M. and Donald, A.M. (1997). Use of the Environmental Scanning Electron Microscope for the Observation of the Swelling Behaviour of Cellulosic Fibres. *Scanning*, **19**, 92–97.

Jenkins, L.M. and Donald, A.M. (1999). Contact Angle Measurements on Fibres in the Environmental Scanning Electron Microscope. *Langmuir*, **15**, 7829–7835.

Jenkins, L.M. and Donald, A.M. (2000). Observing fibers swelling in water with an environmental scanning electron microscope. *Textile Res. J.*, **70**(3), 269–276.

Kaegi, R. and Holzer, L. (2003). Transfer of a single particle for combined ESEM and TEM analyses. *Atmos. Env.*, **37**(31), 4353–4359.

Kalceff, M.A. (2002). Residual surface potentials in the variable pressure environmental scanning electron microscope. *Microsc. Microanal.*, **8**(Suppl. 2), 1534–1535.

Keddie, J.L., Meredith, P., Jones, R.A.L. and Donald, A.M. (1995). Kinetics of Film Formation in Acrylic Latices Studied with Multiple-Angle-of-Incidence Ellipsometry and Environmental SEM. *Macromolecules*, **28**(8), 2673–2682.

Keddie, J.L., Meredith, P., Jones, R.A.L. and Donald, A.M. (1996a). Film Formation of Acrylic Latices with Varying Concentrations of Non-Film-Forming Latex Particles. *Langmuir*, **12**(16), 3793–3801.

Keddie, J.L., Meredith, P., Jones, R.A.L. and Donald, A.M. (Eds) (1996b). *Rate-Limiting Steps in Film Formation of Acrylic Latices as Elucidated with Ellipsometry and Environmental Scanning Electron Microscopy*. ACS Symposium Series 648 Film Formation in Waterborne Coatings, American Chemical Society.

Khouchaf, L. and Boinski, F. (2007). Environmental Scanning Electron Microscope study of SiO_2 heterogeneous material with helium and water vapor. *Vacuum*, **81**(5), 599–603.

Khouchaf, L. and Verstraete, J. (2007). Multi-technique and multi-scale approach applied to study the structural behavior of heterogeneous materials: natural SiO_2 case. *J. Mat. Sci.*, **42**(7), 2455–2462.

Klemenso, T., Appel, C.C. and Mogensen, M. (2006). *In situ* observations of microstructural changes in SOFC anodes during redox cycling. *Electrochem. Solid State Lett.*, **9**(9), A403–A407.

Kucheyev, S.O., Toth, M., Baumann, T.F., Hamza, A.V., Ilavsky, J., Knowles, W.R., Saw, C.K., Thiel, B.L., Tileli, V., van Buuren, T., Wang, Y.M. and Willey, T.M. (2007). Structure of low-density nanoporous dielectrics revealed by low-vacuum electron microscopy and small-angle X-ray scattering. *Langmuir*, **23**(2), 353–356.

Laskin, A., Iedema, M.J., Ichkovich, A., Graber, E.R., Taraniuk, I. and Rudich, Y. (2005). Direct observation of completely processed calcium carbonate dust particles. *Faraday Discuss.*, **130**, 453–468.

Lauri, A., Riipinen, I., Ketoja, J.A., Vehkamaki, H. and Kulmala, M. (2006). Theoretical and Experimental Study on Phase Transitions and Mass Fluxes of Supersaturated Water Vapor onto Different Insoluble Flat Surfaces. *Langmuir*, **22**(24), 10061–10065.

Liukkonen, A. (1997). Contact angle of water on paper components: sessile drops versus environmental scanning electron microscope measurements. *Scanning*, **19**, 411–415.

Lux, A., Luxova, M., Abe, J., Morita, S. and Inanaga, S. (2003). Silicification of bamboo (*Phyllostachys heterocycla Mitf.*) root and leaf. *Plant And Soil*, **255**(1), 85–91.

Ma, Y.R., Borner, H.G., Hartmann, J. and Golfen, H. (2006). Synthesis of DL-alanine hollow tubes and core-shell mesostructures. *Chemistry - A Eur. J.*, **12**(30), 7882–7888.

MacKinlay, K.J., Allison, F.J., Scotchford, C.A., Grant, D.M., Oliver, J.M., King, J.R., Wood, J.V. and Brown, P.D. (2004). Comparison of environmental scanning electron microscopy with high vacuum scanning electron micrcscopy as applied to the assessment of cell morphology. *J. Biomed. Mat. Res.*, **69**, 359–366.

Madsen, D.N., Molhave, K., Mateiu, R., Boggild, P., Rasmussen, A.-M., Appel, C.C., Brorson, M. and Jacobsen, C.J.H. (2003). *Nanoscale soldering of positioned carbon nanotubes using highly conductive electron beam induced gold deposition*. Third IEEE Conference on Nanotechnology.

Meredith, P. and Donald, A.M. (1996). Study of 'wet' polymer latex systems in environmental scanning electron microscopy: some imaging considerations. *J. Microsc.*, **181**(pt.1), 23–35.

Meredith, P., Donald, A.M. and Payne, R.S. (1996). Freeze-Drying: *In Situ* Observations using Cryoenvironmental Scanning Electron Microscopy and Differential Scanning Calorimetry. *J. Pharmaceut. Sci.*, **85**(6), 631–637.

Micic, M., Jeremic, M., Radotic, K., Mavers, M. and Leblanc, R.M. (2000). Visualisation of artificial lignin supramolecular structure. *Scanning*, **22**, 288–294.

Micic, M., Zheng, Y.J., Moy, V., Zhang, X.H., Andreopoulos, F.M. and Leblanc, R.M. (2003). Comparative studies of surface topography and mechanical properties of a new, photo-switchable PEG-based hydrogel. *Colloids Surf. B – Biointerfaces*, **27**(2–3), 147–158.

Misirli, Z., Oner, E.T. and Kirdar, B. (2007). Real imaging and size values of *Saccharomyces cerevisiae* cells with comparable contrast tuning to two environmental scanning electron microscopy modes. *Scanning*, **29**(1), 11–19.

Molhave, K., Madsen, D.N., Rasmussen, A.-M., Carlsson, A., Appel, C.C., Brorson, M., Jacobsen, C.J.H. and Boggild, P. (2003). Solid gold nanostructures fabricated by electron beam deposition. *Nano Lett.*, **3**(11), 1499–1503.

Multi-authors (2004). Special issue: Characterisation of Nonconductive Materials. *Microsc. Microanal.*, **10**(6).

Muscariello, L., Rosso, F., Marino, G., Giordano, A., Barbarisi, M., Cafiero, G. and Barbarisi, A. (2005). A critical review of ESEM applications in the biological field. *J. Cellular Phys.*, **205**, 328–334.

Myers, B.D. and Dravid, V.P. (2006). Variable pressure electron beam lithography (VP-eBL): A new tool for direct patterning of nanometer-scale features on substrates with low electrical conductivity. *Nano Lett.*, **6**(5), 963–968.

Neuhaus, W., Lauer, R., Oelzant, S., Fringeli, U.P., Ecker, G.F. and Noe, C.R. (2006). A novel flow based hollow-fiber blood-brain barrier in vitro model with immortalised cell line PBMEC/C1-2. *J. Biotechnol.*, **125**(1), 127–141.

Niitsuma, J., Sekiguchi, T., Yuan, X.L. and Awano, Y. (2007). Electron beam nanoprocessing of a carbon nanotube film using a variable pressure scanning electron microscope. *J. Nanosci. Nanotechnol.*, **7**(7), 2356–2360.

Niitsuma, J., Yuan, X.L., Ito, S. and Sekiguchi, T. (2005). Processing of carbon nanotubes with electron beams in gas atmospheres. *Scripta Materialia*, **53**(6), 703–705.

Niitsuma, J., Yuan, X.L., Koizumi, S. and Sekiguchi, T. (2006). Nanoprocessing of diamond using a variable pressure scanning electron microscope. *Japan. J. Appl. Phys. Part 2 – Lett. Express Lett.*, **45**(1–3), L71–L73.

Phillips, M.R., Toth, M. and Drouin, D. (1999). Depletion layer imaging using a gaseous secondary electron detector in an environmental scanning electron microscope. *Appl. Phys. Lett.*, **75**(1), 76–78.

Plucknett, K.P., Baker, F.S., Normand, V. and Donald, A.M. (2001). Fractographic examination of hydrated biopolymer gel composites using environmental scanning electron microscopy. *J. Mat. Sci. Lett.*, **20**(16), 1553–1557.

Pooley, G.D. (2004). Secondary and backscattered electron imaging of weathered chromian spinel. *Scanning*, **26**, 240–249.

Postek, M.T. and Vladar, A.E. (2004). New applications of variable-pressure/ environmental microscopy to semiconductor inspection and metrology. *Scanning*, 26, 11–17.

Rizzieri, R., Baker, F.S. and Donald, A.M. (2003a). A study of the large strain deformation and failure behaviour of mixed biopolymer gels via *in situ* ESEM. *Polymer*, 44(19), 5927–5935.

Rizzieri, R., Baker, F.S. and Donald, A.M. (2003b). A tensometer to study strain deformation and failure behavior of hydrated systems via *in situ* environmental scanning electron microscopy. *Rev. Scientific Instr.*, 74(10), 4423–4428.

Roberts, R.A., Shukla, A.J. and Rice, T. (1997). Characterisation of Polyox granules using environmental scanning electron microscopy. *Scanning*, 19, 104–108.

Rodriguez-Navarro, C., Doehne, E. and Sebastian, E. (2000). Influencing crystallization damage in porous materials through the use of surfactants: Experimental results using sodium dodecyl sulfate and cetyldimethylbenzylammonium chloride. *Langmuir*, 16(3), 947–954.

Rossi, M.P., Ye, H.H., Gogotsi, Y., Babu, S., Ndungu, P. and Bradley, J.C. (2004). Environmental scanning electron microscopy study of water in carbon nanopipes. *Nano Lett.*, 4(5), 989–993.

Royall, C.P., Stokes, D.J., Hopkinson, I. and Donald, A.M. (2001). Confocal Microscopy and Environmental SEM for Polymers. *Polym. News*, 26(7), 226–233.

Samara-Ratna, P., Atkinson, H.V., Stevenson, T., Hainsworth, S.V. and Sykes, J. (2007). Design of a micromanipulation system for high temperature operation in an environmental scanning electron microscope (ESEM). *J. Micromech. Microeng.*, 17(1), 104–114.

Samatham, R., Park, I.S., Kim, K.J., Nam, J.D., Whisman, N. and Adams, J. (2006). Electrospun nanoscale polyacrylonitrile artificial muscle. *Smart Mat. Struct.*, 15(6), N152–N156.

Saville, P.A., Hainsworth, S.V. and Rutty, G.N. (2007). Cutting crime: the analysis of the "uniqueness" of saw marks on bone. *Int. J. Legal Med.*, 121(5), 349–357.

Shah, H.J., Fontecchio, A.K., Rossi, M.P., Mattia, D. and Gogotsi, Y. (2006). Imaging of liquid crystals confined in carbon nanopipes. *Appl. Phys. Lett.*, 89(4).

Singler, T.J., Clum, J.A. and Prack, E.R. (1993). Dynamics Of Soldering Reactions – Microscopic Observations. *Microsc. Res. Techn.*, 25(5–6), 509–517.

Smith, A.J., Atkinson, H.V., Hainsworth, S.V. and Cocks, A.C.F. (2006). Use of a micromanipulator at high temperature in an environmental scanning electron microscope to apply force during the sintering of copper particles. *Scripta Materialia*, 55(8), 707–710.

Sorensen, B.F., Horsewell, A. and Skov-Hansen, P. (2002). *In situ* observations of crack formation in multi-filament Bi-2223HTS tapes. *Physica C – Supercond. Appl.*, 372, 1032–1035.

Sorgi, C. and De Gennaro, V. (2007). ESEM analysis of chalk microstructure submitted to hydromechanical loading, *Comptes Rendus Geoscience*, 339(7), 468–481.

Stelmashenko, N.A., Craven, J.P., Donald, A.M., Terentjev, E. and Thiel, B.L. (2001). Topographic Contrast of Partially Wetting Water Droplets in Environmental Scanning Electron Microscopy. *J. Microsc.*, 104(2), 172–183.

Stelmashenko, N. and Donald, A.M. (1998). *ESEM Study of Film Formation in Latices Polymerised in Presence of Starch*. Microscopy & Microanalysis '98, Atlanta, GA.

Stokes, D.J. (2006). Progress in the study of biological specimens using ESEM. *In focus* (Proceedings of the RMS), 2(June 2006), 64–72.

Stokes, D.J., Baker, F.S. and Toth, M. (2004a). Raising the pressure: realising room temperature/high humidity applications in ESEM. *Microsc. Microanal.*, 10(Suppl. 2), 1074–1075.

Stokes, D.J. and Donald, A.M. (2000). *In Situ* Mechanical Testing of Dry and Hydrated Breadcrumb using Environmental SEM. *J. Mat. Sci.*, 35, 599–607.

Stokes, D.J., Mugnier, J.Y. and Clarke, C.J. (2004b). Static and dynamic experiments in cryo-electron microscopy: comparative observations using high-vacuum, low-voltage and low-vacuum SEM. *J. Microsc. – Oxford*, 213, 198–204.

Stokes, D.J., Rea, S.M., Best, S.M. and Bonfield, W. (2002). Electron Microscopy of Human Osteoblasts in the Absence of Fixing, Drying, Freezing or Specimen Coating, *Scanning*, 25(4), 181–184.

Stokes, D.J., Thiel, B.L. and Donald, A.M. (1998). Direct Observations of Water/Oil Emulsion Systems in the Liquid State by Environmental Scanning Electron Microscopy. *Langmuir*, 14(16), 4402–4408.

Stokes, D.J., Thiel, B.L. and Donald, A.M. (2000). Dynamic Secondary Electron Contrast Effects in Liquid Systems Studied by Environmental SEM (ESEM). *Scanning*, 22(6), 357–365.

Tai, S.S.W. and Tang, X.M. (2001). Manipulating Biological Samples for Environmental Scanning Electron Microscopy Observation. *Scanning*, 23, 267–272.

Thaveeprungsriporn, V., Mansfield, J.F. and Was, G.S. (1994). Development Of An Economical Electron Backscattering Diffraction System For An Environmental Scanning Electron-Microscope. *J. Mat. Res.*, 9(7), 1887–1894.

Thiel, B.L. and Donald, A.M. (1998). *In Situ* Mechanical Testing of Fully Hydrated Carrots (*Daucus carota*) in the Environmental SEM. *Ann. Bot.*, 82, 727–733.

Thiel, B.L., Toth, M., Schroemges, R.P.M., Scholtz, J.J., van Veen, G. and Knowles, W.R. (2006). Two-stage gas amplifier for ultrahigh resolution low vacuum scanning electron microscopy. *Rev. Scientific Instr.*, 77(3).

Timofeeff, M.N., Lowenstein, T.K. and Blackburn, W.H. (2000). ESEM-EDS: an improved technique for major element chemical analysis of fluid inclusions. *Chem. Geol.*, 164(3–4), 171–182.

Toth, M. and Baker, F.S. (2004). Secondary Electron Imaging at Gas Pressures in Excess of 15 torr. *Microsc. Microanal.*, 10(Suppl. 2), 1062–1063.

Toth, M., Knowles, W.R. and Thiel, B.L. (2006). Secondary electron imaging of nonconductors with nanometer resolution. *Appl. Phys. Lett.*, 88(2), Article no. 023105.

Toth, M., Lobo, C.J., Hartigan, G. and Knowles, W.R. (2007a). Electron flux controlled switching between electron beam induced etching and deposition. *J. Appl. Phys.*, 101(5).

Toth, M., Lobo, C.J., Knowles, W.R., Phillips, M.R., Postek, M.T. and Vladar, A.E. (2007b). Nanostructure fabrication by ultra-high-resolution environmental scanning electron microscopy. *Nano Lett.*, 7(2), 525–530.

Toth, M. and Phillips, M.R. (2000). The role of induced contrast on images obtained using the environmental scanning electron microscope. *Scanning*, 22, 370–379.

Toth, M., Thiel, B.L., Coy, M.A., Marshman, J.G. and Knowles, W.R. (2005). Artifact-free imaging of Photolithographic Masks by Environmental Scanning Electron Microscopy. *Microsc. Microanal.*, **11**(Suppl. 2), 390–391.

Toth, M., Uncovsky, M., Knowles, W.R. and Baker, F.S. (2007c). Secondary electron imaging at gas pressures in excess of 1kPa. *Appl. Phys. Lett.*, **91**, Article No. 053122.

Tricart, J.-P., Durrieu, J. and Lagarde, F. (1997). Imaging of Pseudo Oil Base Mud by Environmental Scanning Electron Microscopy. *Rev. L'Institute Francais Petrole*, **52**(2), 151–159.

Verstraete, J., Khouchaf, L. and Tuilier, M.H. (2004). Contributions of the environmental scanning electron microscope and X-ray diffraction in investigating the structural evolution of a SiO_2 aggregate attacked by alkali-silica reaction. *J. Mat. Sci.*, **39**(20), 6221–6226.

Wako, I., Chokan, T., Takagi, D., Chiashi, S. and Homma, Y. (2007). Direct observation of single-walled carbon nanotube growth processes on SiO_2 substrate by *in situ* scanning electron microscopy. *Chem. Phys. Lett.*, **449**(4–6), 309–313.

Waller, D. (2007). *Environmental scanning electron microscopy of freezing aqueous solutions*, University of Cambridge.

Waller, D., Donald, A.M. and Stokes, D.J. (2008). Improvements to a cryo-system to observe ice nucleating in a variable pressure SEM. *Rev. Sci. Inst.* (accepted).

Warner, M., Thiel, B.L. and Donald, A.M. (2000). The Elasticity and Failure of Fluid-Filled Cellular Solids: Theory and Experiment. *Proc. Natl Acad. Sci.*, **97**(4), 1370–1375.

Wei, Q.F., Wang, X.Q., Mather, R.R. and Fotheringham, A.F. (2003). ESEM study of size removal from ceramic fibers by plasma treatment. *Appl. Surf. Sci.*, **220**(1–4), 217–223.

Weimer, P.J., Price, N.P.J., Kroukamp, O., Joubert, L.-M., Wolfaardt, G.M. and Van Zyl, W.H. (2006). Studies of the extracellular glycocalyx of the anaerobic cellulolytic bacterium *Ruminococcus albus* 7. *Appl. Environ. Microbiol.*, **72**(12), 7559–7566.

White, R., Phillips, M.R., Thomas, P. and Wuhrer, R. (2006). In-situ investigation of discolouration processes between historic oil paint pigments. *Microchim. Acta*, **155**(1–2), 319–322.

Williams, S.J., Morrison, D.E., Thiel, B.L. and Donald, A.M. (2005). Imaging of semiconducting polymer blend systems using environmental scanning electron microscopy and environmental scanning transmission electron microscopy. *Scanning*, **27**(4), 190–198.

Wirth, C.T., Hofmann, S. and Robertson, J. (2008). Surface properties of vertically aligned carbon nanotube arrays. *Diamond Related Mat.* (in press).

Xiao, J.Z., Shao, M.J. and Yin, S.T. (2002). Environmental SEM investigation on surface defects in 0.92Pb(Zn1/3Nb2/3)O-3-0.08PbTiO(3) single crystal. *J. Crystal Growth*, **240**(3–4), 521–525.

Zhu, S. and Cao, W. (1997). Direct observation of ferroelectric domains in $LiTaO_3$ using environmental scanning electron microscopy. *Appl. Phys. Lett.*, **79**, 2558–2561.

Zimmermann, F., Ebert, M., Worringen, A., Schutz, L. and Weinbruch, S. (2007). Environmental scanning electron microscopy (ESEM) as a new technique to determine the ice nucleation capability of individual atmospheric aerosol particles. *Atmos. Env.*, **41**(37), 8219–8227.

Index

Printed and bound by CPI Group (UK) Ltd, Croydon, CR0 4YY

27/10/2024

14580157-0001